TIME AND REVOLUTION

Stephen E. Hanson

TIME

AND REVOLUTION

Marxism and the Design of Soviet Institutions

The
University
of North
Carolina
Press
Chapel Hill
& London

HM
73
H316
1997

Manufactured in the United States of America
The paper in this book meets the guidelines for permanence and
durability of the Committee on Production Guidelines for Book
Longevity of the Council on Library Resources.

Library of Congress Cataloging-in-Publication Data
Hanson, Stephen, 1963–
Time and revolution: Marxism and the design of Soviet
institutions / Stephen E. Hanson.
    p.   cm.
Includes bibliographical references and index.
ISBN 0-8078-2305-8 (cloth: alk. paper).—
ISBN 0-8078-4615-5 (pbk.: alk. paper)
1. Time—Social aspects. 2. Revolutions and socialism. 3. Soviet
Union—Economic policy. 4. Statics and dynamics (Social sciences).
5. Time and economic reactions. I. Title.
HM73.H316   1997   96-13723
304.2'3—dc20   CIP

01 00 99 98 97   5   4   3   2   1

Chapter 6 was previously published, in a slightly different form,
as "Gorbachev: The Last True Leninist Believer?," in *The Crisis of
Leninism and the Decline of the Left*, ed. Daniel Chirot (Seattle:
University of Washington Press, 1991).

Contents

PREFACE

he clock was accurate, but Margulies did not depend on it. He was not asleep. He always rose at six and was always ahead of time. There had never yet been an occasion when the alarm clock had actually awakened him.

Margulies could not really have faith in so simple a mechanism as a timepiece; could not entrust to it so precious a thing as time.
—*Valentin Kataev,* Time, Forward!, *1932*

The passage above, which begins a prominent Stalinist propaganda novel extolling the virtues of heroic efforts to overfulfill the First Five-Year Plan, unwittingly encapsulates a central dilemma of Soviet development: how might a regime whose purpose was to build a communist society beyond the constraints of ordinary time nonetheless enforce economically productive norms of time use so as to compete with the capitalist West?

Margulies, the archetypal Bolshevik engineer of the novel, is a character unimaginable in a Western context. Certainly, the idea that time is too precious to entrust to a watch seems hard to fathom from a capitalist point of view. What is time, anyway, besides the name we give to what a watch measures? A capitalist engineer reading Kataev's book today would simply laugh at its main conclusion—that economic productivity under socialism could have been enhanced if everybody in the Soviet Union had been like Margulies and thrown away his or her watch so as to utilize time in a more revolutionary manner.

Yet, as this study will show, Stalinist economic institutions were originally set up to encourage precisely this sort of behavior. Many of the well-known structural problems of the Soviet planned economy—the enormous waste of human and material resources, the tendency of factories to alternate between periods of frantic rush work and periods of prolonged inaction, the shoddiness of goods, and the lack of incentives to innovate—become explicable once we take into account something often overlooked in Western analyses: the men who designed this economy were not interested in trying to achieve efficiency in the Western sense.[1] If the ideal of "bourgeois" economists is a system in which each unit of time is utilized in as productive a manner as possible given scarce resources, the goal of Soviet socialism was to organize production in such a manner as to master time itself.

Early liberal capitalist regimes forced newly urbanized workers to adjust to the idea that the rule of abstract time is inexorable; work must therefore be steady and disciplined and "free time" kept within strict bounds. The Soviet regime held out a different promise: that if work was done intensively enough, according to the party's direction, time could actually be compressed and the conflict between labor and leisure ultimately overcome.[2] The institutionalization of this idea created a sort of final-exam economy—since an endless summer vacation (communism) was always held to be just around the corner, the most rational thing to do was to "cram" (or "storm") so as not to have to repeat the course.[3] Thus, the Stalinist economy temporarily mobilized workers for intense, even frenzied activity but did not provide any basis for the routinization of that activity according to norms of modern time discipline.[4] And once Stalinist methods of mobilization were abandoned, the basic irrationalities of the design of that economy led inexorably to its downfall.

But why was such a strange system of economic incentives, from the Western point of view, established in the first place? This book shows that the characteristic inefficiencies of time use under Stalinist economics can be traced ultimately to the ideological framework developed by Karl Marx—in particular, to Marx's conception of time itself. Without a careful analysis of this quite novel conception of time and its historical development from Marx to Lenin and Stalin, I argue, the specific design of Soviet economic institutions must remain incomprehensible. Moreover, in the absence of a coherent explanation of where the Soviet system came from, the reasons for its decline and fall in the post-Stalin era must also remain obscure.

Following Ken Jowitt's application of the Weberian types of "legitimate domination" to Leninist regimes, I term the distinctive view of time originated in the works of Marx and institutionalized under Lenin and Stalin the *charismatic-rational conception of time.*[5] Marxism, Leninism, and Stalinism, I will show, all rejected the ideal-typically "rational" conception of time as an abstract, linear grid outside all concrete events—the time orientation that is politically and economically institutionalized and, to a remarkable degree, culturally internalized in the liberal capitalist West. At the same time, they also rejected the purely "charismatic" conception of time as a force that can be transcended through revolutionary will—the time orientation characteristic of, for example, revolutionary anarchists. Instead, the ideology ultimately codified as "Marxism-Leninism" was based on Marx's principle that effective revolutionary praxis depends upon utilizing rational time discipline to master time itself.

I argue that the development and institutionalization of Marx's charismatic-rational conception of time occurred in three broad cycles. The first, which I term the *theoretical cycle*, begins with the development of Marx's synthesis of scientific analysis and revolutionary action in his theory of historical materialism; this cycle ends with the collapse of the socialist Second International during World War I. The second period—the *political cycle*—begins with the seizure of state power by the Leninist "party of professional revolutionaries" in November 1917 and ends with the destruction of the New Economic Policy (NEP) in 1928. Finally, the *socioeconomic cycle* begins with Stalin's synthesis of "American efficiency" and "Russian revolutionary sweep" in the institutions of the Five-Year Plan and ends with the "period of stagnation" under Brezhnev.

It should be emphasized that the unfolding of these cycles was by no means inevitable. Indeed, throughout the historical period under examination here, the problem of how to institutionalize Marx's vision of socialism was the source of continual controversy and factional rivalry. Had these controversies been resolved differently, Marxism might never have generated anything like the Soviet system. But the precise form the historical debates among Marxists and Leninists took, I will argue, can be fully understood only in terms of the charismatic-rational conception of time underlying Marxist discourse, for in each cycle—the theoretical, the political, and the socioeconomic—a remarkably similar pattern occurred. First, a revolutionary innovator created new formal institutions designed to synthesize charismatic time transcendence and modern time discipline. Then, after the death of that innovator, the elite split into three camps: a "left" group that adopted an essentially charismatic time orientation, a "right" group that adopted an essentially rational time orientation, and a "center" group that dedicated itself to maintaining orthodoxy against attacks from both left and right. In each cycle, these splits led to a decline in revolutionary momentum and a danger of the movement's collapse.

Adopting this theoretical approach, we can see the entire history of Marxism and of the Soviet Union, from the writing of the *Communist Manifesto* to the disintegration of the Soviet bloc, as constituting an unprecedented, often highly coercive, and ultimately unsuccessful 150-year revolutionary experiment in reordering the human relationship to time—an experiment whose history from beginning to end displayed a remarkably consistent developmental logic. Moreover, if this approach is correct, failure to take into account the distinctive nature of the revolutionary project linking Marx to Gorbachev will be a serious obstacle to making sense of what emerges from the rubble of that project in the years to come.

I have divided the present work into six chapters. Chapter 1 examines the social science literature dealing with time conceptions and time discipline in various social contexts. After providing a critique of this literature, I set out an alternative Weberian theory of social time that distinguishes between types of organization based upon three basic time conceptions: the traditional, the rational, and the charismatic. This framework is then utilized to contrast the institutionalization of rational time under liberal capitalism in the West with the organization of social time under Leninism.

Chapter 2 examines the problem of time in the works of Kant and Hegel, which are critical for understanding Marxist thought. The conceptions of time developed by these philosophers are interpreted as a reaction to, and partial rejection of, the triumph of rational time in western Europe. The chapter concludes with a brief examination of philosophical debates within the Hegelian school after Hegel's death.

Chapter 3 focuses on the first, theoretical cycle of development in the history of Marxism, from the publication of the *Communist Manifesto* in 1848 to the destruction of the Second International during World War I. I examine in detail the problem of time in the theoretical work of

Karl Marx, arguing that Marx's work presents a novel, if not wholly consistent, interpretation of the nature of time—labor time in particular—that can be interpreted in Weberian terms as an amalgam of charismatic and rational time conceptions. This is followed by an analysis of the "routinization" of revolutionary Marxism manifested in the works of Bernstein, Luxemburg, and Kautsky.

Chapter 4 examines the second, political cycle in the history of charismatic-rational time, focusing on the institutional evolution and eventual political triumph of Lenin's "party of professional revolutionaries"—itself an innovative synthesis of charisma and rationality in the political organization of time. I then turn to a discussion of the decline of political Leninism as Trotsky, Bukharin, Zinoviev, and Stalin struggled for power in the 1920s.

Chapter 5 analyzes the socioeconomic institutions built during Stalin's First Five-Year Plan as the beginning of a third cycle of development. I argue that these institutions are based on a charismatic-rational form of "planned heroism" that can also be understood as emerging out of the conception of time developed by Marx and Lenin—now on the economic level. This chapter concludes with a discussion of the period of decline of socioeconomic Stalinism under Malenkov, Khrushchev, and Brezhnev.

Chapter 6 deals with the period of Gorbachev's *perestroika* from 1985 to 1991. I argue that Gorbachev's reforms were intended not to repudiate the conception of time implicit in Marxist theory and institutionalized in Leninist politics and Stalinist economics but rather to usher in a new period of *cultural* charismatic rationalism in time use—that is, the internalization by Soviet citizens of the specific norms of "revolutionary time discipline" envisioned by Marx, Lenin, and Stalin. I trace how Gorbachev's pursuit of this goal led not, as he had intended, to the beginning of a new phase of Leninist development but instead to the collapse of the Stalinist economy and the Leninist polity.

The conclusion presents an overview of the rise and fall of the Soviet Union from the theoretical perspective developed throughout the book, assessing in a preliminary manner the degree to which the formal institutions of the Soviet regime succeeded in transforming the everyday time sense of the social groups subjected to them. I then briefly suggest how analyzing the history of the Leninist charismatic-rational conception of time can enrich our understanding of political, socioeconomic, and cultural development in the West and elsewhere.

Considering the scope of this work, a few words of caution are in order. It should be emphasized at the outset that this book is meant only to outline a novel interpretation of Leninist rule based on the Soviet case; it is by no means intended as a comprehensive social history of time use in prerevolutionary Russia or the USSR. Specifically, the focus throughout is on elite struggles to define and institutionalize socialism, struggles in which conceptions of time played a central part. I emphasize the role of elites not because identifying the sources of broader social support for and resistance to Leninist policies is somehow unimportant in explaining the course of Soviet history; clearly, the implementation of the formal institutions favored by Lenin and Stalin could not have taken place without some degree of social support for them, nor would they have evolved in the same manner had there not been significant social resistance to their implementation. However, I am primarily interested here in accounting for the formal design of the political and socioeconomic institutions within which social struggles in the Soviet Union took place, and this requires a detailed examination of the motivations of, and debates among, the designers. To analyze the diverse reactions of different sectors of Russian society to Bolshevik elite policies in different periods of Soviet history—as would be necessary in any study of the organization of "social time" in the Soviet Union more generally—would require a whole series of separate works. Hopefully, the necessarily "top down" analysis of Leninist elites presented here will prove complementary to investigations of everyday time use in Soviet society from the "bottom up."[6]

In addition, although this analysis is embedded within a comparative theoretical framework, it remains essentially a single case study. No attempt is made to apply the framework developed here to other Leninist regimes such as China, North Korea, or Cuba, although further research into the generalizability of the argument would be of great interest. However, given the fact that the Soviet Union was the first and, until 1989, the most powerful Leninist regime, explaining the origins and inherent contradictions of its institutional design remains in any case of central interest for the study of communism and postcommunism.

Finally, to argue that ideology mattered to Leninist elites throughout the Soviet period—let alone that it played the central role I ascribe to it here—admittedly contradicts the general tendency of social scientists to downplay the independent influence of ideas in history.[7] It is my conviction that a large part of the reason for scholarly skepticism about the causal role of ideology is that few analyses endeavor to show concretely, rather than speculatively, how ideological visions get translated into institutional outcomes. Hopefully, then, this book will inspire a reevaluation of the theoretical marginalization of the study of ideology's effects on institutional order—in the Soviet case and elsewhere.

ACKNOWLEDGMENTS

This book could not have been written without the support of a great many people, not all of whom I will be able to mention here. The good things in this work I owe chiefly to their teaching, criticism, and advice during the book-writing process; the bad things are, of course, my own responsibility.

First and foremost, I would like to thank Ken Jowitt, who has been an inspiration to me intellectually and personally for many years. My debt to Ken is enormous; there is no way I could possibly have made any sense of the Marxist-Leninist conception of time without the benefit of his intellectual framework, and my life has been enriched in every way by our collaboration.

Secondly, I would like to thank Gail Lapidus, George Breslauer, and Gregory Grossman, both for their helpful comments on earlier drafts of this work and for their leadership in founding the Berkeley-Stanford Program on Soviet Studies, which provided the kind of dynamic academic environment in which new approaches to the study of the USSR could truly flourish. I also thank them for their unfailing personal support of my work at Berkeley. In addition, I would like to thank Hanna Pitkin for her comments on my early drafts and for her inspiring teaching of political theory. I hope I have done it justice.

TIME AND REVOLUTION

1
TRADITIONAL, MODERN, AND CHARISMATIC TIME

For people who live in modern liberal capitalist societies, time is an omnipresent force. Popular magazine articles and public administration courses alike stress the importance of proper time management, punctuality is taken as a sign of maturity, and one's first waking moments are often spent staring at the digital display of a buzzing alarm clock. But in our time-saturated culture it is hard to keep in mind that things were not always so. In fact, as recently as the late nineteenth century, the prospect of standardizing time measurement on a national scale aroused active political resistance; two hundred years ago, the vast majority of Europeans worked in an agricultural setting whose cyclically changing demands on human activity allowed for fairly extended periods of relative relaxation; before about 1450, the mechanical clock itself was practically unknown.[1]

The decisive transformation in social time perception accompanying industrialization in the West—conventionally labeled as the change from a traditional time sense to a modern one—has become a standard theme in sociology and comparative history.[2] Many different definitions of traditional and modern time have been proposed, but perhaps the most fundamental distinction between the two conceptions of time is that in traditional cultures, by and large, time is conceived of as concrete, tied to the flow of actual events; in modern societies, by contrast, time appears as abstract, a kind of universal grid against which the duration of all particular events can be measured.[3] The clearest manifestation of this distinction is the shift in the predominant mode of time measurement, from observing the movements of heavenly bodies to the keeping of mechanical clocks. The traditional world was one governed by the sun's position in the sky, the phases of the moon, and the seasons of the year; the modern world relies on recent social inventions such as abstract "seconds" and "minutes."[4]

Related to this distinction between concrete and abstract time is a secondary distinction between cyclical and linear conceptions of time. Since the time measurement of traditional cultures is most intimately connected with the cyclical movements of the heavenly bodies, these cosmic patterns tend to provide a framework for understanding time itself in terms of repeating cycles, and it is the cyclical metaphor for time that is predominant in traditional religion and philosophy. Conversely, once time is seen as an abstract grid outside all concrete events, only the addition of directionality—from the past toward the future—is required to make time appear to be an infinite line, extending perhaps even before and after the period of the universe's existence.[5]

The ethos of a society for which time is synonymous with the flow of concrete events is well captured by Evans-Pritchard in his description of the time sense of the Nuer tribe in East Africa:

> The Nuer have no expression equivalent to "time" in our language, and they cannot, therefore, speak of time as though it were something . . . which passes, can be wasted, can be saved, and so forth. I do not think that they ever experience the same feeling of fighting against time, because their points of reference are mainly the activities themselves, which are generally of a leisurely character. Events follow a logical order, but they are not controlled by an abstract system, there being no autonomous points of reference to which activities have to conform with precision. Nuer are fortunate.[6]

In fact, such a description of the traditional time sense—whether or not one accepts Evans-Pritchard's judgment of its superiority—appears to be equally apt for the vast majority of people living in preindustrial societies. Of course, the traditional understanding of time as concrete does not preclude the development among traditional elites—the ancient Greek philosophers, for example—of highly sophisticated analyses of the nature of time. Even these philosophical works, however, uniformly link "time" to the flow of actual events, even as they debate whether this flow is cyclical, repetitive, illusory, or representative of some deeper unchanging principle.[7] Not until the modern era do we find a widespread acceptance of the idea that time exists altogether separately from both daily activities and cosmic processes. Nor do elites in premodern societies attempt the mass organization of human activity to conform to purely abstract time demarcations.

The consequences of time orientation for the social order are immense and affect practically every aspect of social life. Some sense for the importance of distinguishing between the time sense predominant in different social contexts can be illustrated by a brief comparison of traditional and modern cultures on two issues intimately bound up with the question of time orientation: the issue of the temporality of the human life span, and the issue of how daily activity is organized in time.

The conception of time as concrete and cyclical predominant in non-Western cultures is associated with a notion of the human relationship to cycles of nature quite different from that typical of the modern West. Birth and death in traditional cultures tend to be seen not as the finite endpoints of a life span conceived in linear terms but instead as passages in a natural process of generational renewal. The concept of childhood as a special phase of an individual's life span is essentially a modern one; traditional societies tend to see life in terms of the cycle of youth, maturity, and old age, with "youth" extending from infancy to young adulthood.[8] Death, as well, is perceived not as the ultimate cessation of existence but as another necessary (if not particularly welcome!) part of the cyclical journey of the human soul, and thus it can be accepted relatively peacefully. As Ariès puts it, "The tame death is the oldest death there is."[9]

The contrast here with the cultural patterns of modern societies is striking. Birth and death are now seen as the absolute endpoints of an individual's life span rather than as passages in a continuing, collective cyclical process. Along with this change has come a new, peculiarly modern anxiety about "wasting one's time"—if one has only, say, seventy years to make one's mark on a time span of history extending, in principle, infinitely into the past and future, one can hardly afford to "lose" any time. The impact of such anxiety on family life, predictably enough, has been enormous. Children are now often subjected to achievement pressures almost immediately after birth; finding the time to nurture intimate relationships becomes increasingly difficult. Finally, the modern attitude toward death is characterized by an almost pathological avoidance of the issue.[10]

Tied to the traditional conception of time as concrete and cyclical is a particular mode of time use in the daily activity of traditional cultures, one that is based on a conception of labor activity and leisure activity as belonging to completely different temporal spheres. Specifically, labor is defined as activity that is necessary for the simple maintenance of human life in the realm of nature governed by cyclical time—the obtaining of food and shelter, the raising of children, and the like.[11] Conversely, leisure is defined as activity that takes place instead in the realm of *freedom from time*.[12] This distinction could take the form of a class distinction between the laboring classes and a leisure aristocracy that abstained from all manual labor, as in ancient Greece. Alternatively, the distinction between labor and leisure could be determined by the consecration of holidays ("holy days"), which were seen as removed from mundane time, when all worldly activity ceased. The Jewish Sabbath is one well-known example of this, but the pattern of associating work days with profane time and holidays with sacred time is common among traditional societies.[13]

As with the traditional conception of time, the modern view of time as abstract and linear has vital implications for how labor and leisure are understood and organized in daily patterns of time use. Within an abstract time frame, the distinction between labor and leisure can be understood only as a distinction between "work time" and "free time"; that is, the distinction is no longer between activity performed under the constraints of time versus activity performed in a state of freedom from time but instead refers merely to activity performed at different times of the day. Holidays, therefore, lose their sacred character in modern society; they are no longer "holy days." Instead, they can be rescheduled or even combined to allow for more convenient vacations, as with the United States holiday "Presidents' Day" in February, which substitutes for the concrete birthdays of *both* Washington and Lincoln in order to create a more efficient three-day weekend. Modern time thus has the advantage of allowing for periodic breaks in activity without losing sight of worldly concerns of efficiency and productivity; however, modern societies pay the price of a marked devaluation of leisure and increased feelings of stress even during what is supposed to be "free time."[14]

Theories of Social Time Orientation

As this overview shows, the analysis of changing conceptions of time can be applied to a wide range of issues of fundamental social importance. An attempt to develop a theory of the relationship between social understandings of time and types of social order, therefore, would seem to be well worth pursuing. However, despite the vast literature on the problems surrounding "time," most treatments of the topic have been primarily philosophical, historical, or interpretive and are not framed in terms of the broader concerns of social science theory.[15] Even fewer studies attempt to explain changes in social time, rather than simply describe them. In order both to clarify and to provide a critique of the competing theories that might explain the dramatic shift from traditional to modern time, one must to some extent tease out the implicit theoretical assumptions of the various authors working on the topic. In general, explanations of changes in time conceptions in the modern capitalist West separate into three basic approaches: modernization theory, neo-Marxism, and cultural relativism.

The modernization perspective on time orientation is perhaps most clearly presented in the work of Inkeles and Smith.[16] Drawing on a tradition of social analysis derived from the work of Talcott Parsons, Inkeles and Smith argue that the shift from traditional to modern time is part of a general unilinear, universal process of rationalization of culture, driven by the rationalization of material life. In other words, they see the changes in technology brought about during industrialization as bringing about concomitant shifts in time perception:

> Industrial production requires precise scheduling in bringing together the diverse elements entering into the production process. This requirement is most evident with the assembly line, since it rigorously imposes the necessity that everyone start and stop at the same time, that each process be allocated a precise amount of time, and that each step be completed as scheduled. According to the socialization principle of *exemplification*, men working in factories should come to internalize a concern for orderly advanced planning and precise scheduling.[17]

From this point of view, liberal capitalist societies have "become modern" in their time orientation essentially because of cumulative cultural exposure to industrial modes of work organization.

This argument is subject to two objections, however. First of all, although it is certainly true that factories that are run according to rationalized time discipline *can* be important "schools of modernity" for those who work in them—since the failure to internalize abstract time norms often results in stiff penalties or dismissal—not all factories are actually run in such a uniform, rationalized way.[18] If factory managers themselves are not convinced of the importance of efficient time use, they are unlikely to organize production so as to bring about any fundamental change in the way their employees use, let alone conceive of, time. This is especially true if the factory in question is protected from the market by state intervention, in which case efficiency may not be crucial for economic survival. Assembly lines by themselves do not preclude traditional afternoon naps or inefficient rush work. Whether a factory becomes an effective means of transmitting a modern time sense, or modern culture in general, thus depends on whether it is run by modern elites with an efficiency orientation who are to some degree reinforced by a peer group with an identifiably modern way of life. As we shall see in the case of Soviet industry, such is by no means automatically the case.

The second objection to Inkeles and Smith's argument concerns their hypothesis that the experience of time in the factory will be generalized to the factory worker's view of time in society as a whole. In fact, it seems quite plausible to expect, at least in some cases, precisely the opposite: to the extent that a factory in a traditional cultural setting successfully approximates modern, abstract norms of time discipline, factory workers who still value traditional, relatively leisurely patterns of time use may be even more opposed to the spread of factory-type work schedules to other areas of social life. Indeed, the persistence of traditional attitudes toward time in many non-Western countries seems quite evident despite the advent of factory work.[19]

A more nuanced version of the modernization argument is presented in David Landes's seminal work *Revolution in Time*. Landes, unlike Inkeles and Smith, explores in detail the crucial role of cultural predispositions in explaining the origins of the primary technological instrument making a modern time orientation possible on a mass scale—the mechanical clock. He argues compellingly that before the technological breakthrough of the use of the pendulum for time measurement could occur, a prior change in thinking about time had to take place; as he puts it, "the clock did not create an interest in time measurement; the interest in time measurement led to the invention of the clock."[20] Landes traces the origins of the interest in modern, abstract time measurement to the medieval Christian monastery, with its regular succession of daily prayers that had to be performed punctually; he contrasts the situation in western Europe with that of China, where the lack of any such religious demand for daily time reckoning led to a concentration of the technologically far more advanced Chinese on the development of ever more sophisticated water clocks, whose cyclical nature betrayed their cultural roots in traditional time perception.

Still, Landes is less persuasive in explaining the rapid spread of modern time reckoning throughout western European society. He first argues that the clock spread from the church to the city because "just like the monastery, the city needed to know the time even before the mechanical clock became available."[21] But the objections given above to the notion that factory work automatically creates the conditions for a cultural change toward modern time perception apply equally to the argument that city life per se creates those conditions. There are, after all, traditional as well as modern cities; ancient Rome prospered for centuries with its daily time measured by sundials. And cities, unlike the individuals who make up social groups, do not *desire* one form of time measurement or another. However, Landes fails to give a causal, as opposed to functionalist, argument to explain why specific city dwellers would have developed a new mode of relating to time at precisely the point in history they did.

At times, in fact, Landes writes as if the invention of the mechanical clock, once it occurred, by itself necessitated the entire subsequent shift in Western time perception: "The fact that earlier usages [of traditional time measurement] persisted well into a new timekeeping era is testimony to the difficulty of changing so fundamental a way of thinking about and ordering life and work. In the long run, however, *change was implicit in the new mode of measurement*."[22] Actually, there is no inherent reason why the introduction of mechanical clocks in a society need bring about an entirely new cultural perception of time as abstract or linear. For example, Landes himself points out that the Japanese ingeniously designed mechanical clocks with expandable faces so that "hours" could vary in length according to seasonal changes in the duration of daylight.[23] Even in present-day liberal capitalist societies, the circular placement of the hours on watch dials themselves has continued to demonstrate the persistence of the perception of time as tied to the concrete cycles of day and night—at least until the very recent introduction of the digital clock. Although the invention of some sort of mechanical clock is quite probably a necessary condition for mass industrialization, since factory labor can scarcely be organized according to sundials or water clocks, it is certainly not by itself a *sufficient* condition for the transition to a decisively modern conception of time throughout society, as Landes implies.

Both Landes's book and that of Inkeles and Smith omit an examination of the case of Leninist industrialization, which might modify their technological determinism. Perhaps more surprisingly, so also does the Marxist historian E. P. Thompson, whose more subtle treatment of changing time perceptions in western Europe nevertheless makes some of the same teleological assumptions as modernization theory.

Compared with the works of modernization theorists, Thompson's account of the origins of Western capitalist time discipline displays a keen sense of the politically contentious nature of time—especially the social organization of labor time and leisure time. Thompson argues persuasively that even in England, the core country of modern capitalism, workers did not fully embrace the timetables of industrial life until several generations had been raised in the new social environment of clock time imposed by capitalist elites. He explores in fascinating detail the perseverance of traditional work patterns in English factories, such as the observance of a day off from work on "Saint Monday," followed by more intense activity later in the week.[24] In resisting the demands of employers for efficiency in time use, workers demonstrated their autonomy and solidarity within the constraints imposed by the capitalist mode of production. Time discipline, in Thompson's account, is analyzed as a mechanism of social control rather than as a neutral indicator of the progress of modernization.

In examining the overall course of development of the modern time sense, however, Thompson reverts to the idea of a necessary teleology in history that must eradicate all premodern, precapitalist forms of culture. He demonstrates the triumph of modern time convincingly in the English case: while earlier generations of English workers fought against the clock itself, later generations fought for legal limitations on the hours of work, thus revealing their fundamental acceptance of the organization of society in terms of abstract time. But Thompson then leaps to the conclusion, without any argument or evidence, that the Soviet-type economy—like all other industrial systems—should also be analyzed as a modernizing force: "Mature industrial societies of all varieties are marked by time-thrift and by a clear demarcation between 'work' and 'life.' . . . Without time discipline, we could not have the insistent energies of industrial man; and whether this discipline comes in the forms of Methodism, or of Stalinism, or of nationalism, it will come to the developing world."[25] Under these assumptions, the important *differences* between the time conceptions central to Methodism, Stalinism, and nationalism, and the possibility that they have quite different effects on the cultural time sense of those societies they dominate, are necessarily missed.

If both Marxist and modernization theories of changing time perceptions tend to be unnecessarily teleological, cultural relativist treatments of different social understandings of time tend to deny any possibility at all of coming to grips with broad cultural change theoretically. An example of such an approach to the study of time can be found in the work of Edward T. Hall.[26] Hall argues that there are eight completely different types of time and that "the rules for understanding one category . . . are not applicable to another category."[27] Furthermore, since different cultures combine these eight different times differently, there can be no overarching theory of social time but only explorations of particular cultural time perceptions. As Hall puts it, "each culture has its own time frame, in which the patterns are unique."[28]

Although Hall's work is nuanced and often quite insightful in its discussion of the different time perspectives of the cultures he has studied, an investigation that begins with the assumption that every culture is unique cannot provide us with any basis for the analysis of grand historical transformations of the type under consideration here. After all, the change to an abstract, linear time grid in western Europe played a crucial part in the rise of the most powerful set of political and economic institutions in human history. Why did this particular view of time generate such power? Where did it come from? What might be the sources of its eventual breakdown? Hall's approach gives us no way to investigate such questions comparatively and systematically.[29]

A Weberian Theory of Social Time

In order both to move beyond the limits of the theoretical frameworks of the authors discussed above and to provide a theoretical basis for the study of changing time perceptions in Leninist regimes, I will adopt the methodological approach of Max Weber as laid out in his major work, *Economy and Society*.[30] It is perhaps rather surprising, given Weber's own interest in the subject, that there has not yet been any attempt to develop a detailed Weberian theory of social time. It is thus necessary briefly to summarize the foundations of Weberian sociology before proceeding.

The theoretical foundation of Weberian analysis is built on two core ideas: methodological individualism and the *verstehen* approach. Methodological individualism, for Weber, does not imply that all, or even most, social actors see themselves as "individuals" rather than as parts of a greater whole. Rather, the term refers to a sociological approach which takes individual action as the basic unit of analysis in investigating larger forms of social organization. In short, for methodological individualists, whether an institution endures or disintegrates depends upon the probability that the actions of the individuals who participate in it remain coordinated in patterns consistent with the institution's formal organization.

The *verstehen* approach to analyzing individual social action is based on the idea that the motivation for such action must be traced ultimately to its meaning for the actor. This means that the social scientist must inevitably be an interpreter, and not only an abstract modeler, of human behavior. It is the *verstehen* approach that distinguishes Weberian methodological individualism from that of the rational choice or neoclassical economic school, since the latter rejects the importance of understanding subjective motivations and instead assumes that all individual action can be analyzed as some variant of utility-maximizing behavior. For Weber, by contrast, the maximization of power, money, or other values—instrumental rationality—is only one of four basic types of social action; in addition, individual behavior can be motivated by value rationality (engaging in behavior defined as valuable regardless of its outcome), affect (behavior motivated by the subjective feeling of an emotional bond with others), and habit (engaging in behavior that is considered customary or is otherwise unquestioned).

In analyzing the origin of social order, the Weberian approach emphasizes the particular importance of value-rational and affectual motivations in inspiring certain individuals to engage in formal institution building. To be sure, Weber did not propose that an emphasis on "irrational" motivations replace analysis of the material interests involved in the process of institutional formation. Rather, Weber argued that the way the pursuit of material *and* ideal interests was institutionalized in a given regime could often itself be traced to the idea systems of the regime's founders: "Not ideas, but material and ideal interests, directly govern men's conduct. Yet very frequently the 'world images' that have been created by 'ideas' have, like switchmen, determined the tracks along which action has been pushed by the dynamic of interest."[31] From this perspective, the supposed dichotomy between the study of "ideas" or "ideology" on the one hand and of "material interests" on the other disappears; instead, the former is seen as the basis for the particular institutional framework in which the latter operates.

A Weberian theory of social time, then, would reject both the technological determinism of modernization theory and Marxism and the pure cultural relativism of many anthropological approaches. Instead, Weberian analysis explains standardized patterns of social time as the consequence of the enforcement of institutional rules about time use by political elites who subjectively feel that the ideological or religious principles underlying these rules are legitimate. Thus, we may expect that the formal conception of time set out in the ideology or religion of institution-building elites will play a crucial role in determining the types of time use considered legitimate in a given social setting.

Weber argues that the diverse forms of political order in human history can be understood in terms of three basic "ideal types" of legitimate domination—the traditional, the rational-legal, and the charismatic. Weber describes the traditional form of legitimate domination as "resting on an established belief in the sanctity of immemorial traditions"; under traditional social organization, "rules which in fact are innovations can be legitimized only by the claim that they have been 'valid of yore,' but have only now been recognized by means of 'Wisdom.'"[32] Conversely, Weber describes legal-rational authority as being based on "a continuous rule-bound conduct of official business"; even "the typical person in authority . . . is himself subject to an impersonal order." Finally, charismatic domination is based upon the claim of a charismatic leader who is perceived by his followers to possess "extraordinary" or superhuman qualities—a claim that the charismatic figure must periodically validate by performing miracles. Charisma is opposed to all established rule systems, whether traditional or rational-legal; because regular economic activity requires precisely such established rules of conduct, Weber argues that charismatic authority is "specifically foreign to economic considerations."[33]

In fact, although Weber himself never analyzes this topic explicitly, each of his three types of legitimate domination can be seen as based upon a prior conceptual understanding of time. *Traditional time*, as has been described above, refers to the sense of duration and connection between events in the absence of abstract temporal demarcations. The traditional type of legitimate domination, based on the sanctity of age-old rules and a hostility to innovation, flows logically from this view of time, since under such a time conception all valid rules must be seen as inherent in the concreteness of time itself. Logically, too, traditional elites can insist that cultural innovation threatens not only the stability of social order but the very destruction of time itself.

Rational (or modern) time is conceived as an abstract grid outside all concrete events. Such a view of time relativizes and thus undermines all particular claims to authority derived from "immemorial traditions," since from a modern point of view all traditions were founded at some point in time and are therefore merely contingent. Rational-legal authority is instead based upon the notion that certain rules of social and political conduct can be seen as inherent in the very unfolding of human development in time; following these abstract, impersonal rules continuously, even if one is at the pinnacle of power, becomes equivalent to accepting the continuous, impersonal nature of time.[34] Conversely, those who cannot internalize rational time discipline—who are continually late for "appointments" or who miss "deadlines"—thereby subvert the rational-legal social order and must be punished.[35]

The third ideal-typical conception of time—*charismatic time*—has not been explicitly analyzed by scholars to date. The charismatic conception of time, however, turns out to be crucial for understanding Marxism and for comparing Soviet political and socioeconomic institutions with those created within the framework of the decisively rational conception of time predominant in the West. From Weber's description it is clear that neither concrete, cyclical time nor abstract, linear time can characterize the charismatic realm. Both cycles and lines are regular and predictable—routinized—and charisma only exists in the unpredictable realm of the extraordinary.

The charismatic view of time might best be understood as the view that *ordinary time is transcended* for those accepting charismatic domination.[36] In its pure form, charismatic leadership requires converts to forgo all worldly considerations to take part in political activity that is seen as reordering the normal course of time—the founding of states, the creation of new religious communities, or the conquest of new territory. It makes sense that charismatic leaders would thrive in periods of social breakdown, when the destruction of daily routines makes it difficult to make sense of time either in terms of abstract rational measurements or through the experience of concrete cycles of repetition. Followers of charismatic leaders in times of crisis obey them, from this point of view, because charismatics demonstrate that they can uniquely and successfully navigate a world where ordinary time no longer appears to function. And once mundane time ceases to have any hold over events, any successful exercise of leadership is likely to be perceived as nothing short of "miraculous."

If charisma is the one type of domination that appears to operate beyond the constraints of ordinary social time, it is understandable why, as Weber puts it, charisma is the "specifically creative revolutionary force in history."[37] By defining new principles of legitimacy in times of social breakdown, charismatic leaders convince their followers that they are, in effect, beginning time anew. It is worth noting, in this context, that seemingly every major revolutionary group in history has attempted either to bring about the end of time or to start time over again—from the Jacobins' revolutionary "Year 1" to the Nazis' "thousand-year reich"; from Marxism's communism at the "end of human prehistory" to Khomeini's references to the apocalyptic "return of the Twelfth Imam."[38]

Analyzing charisma as based upon a collective belief in the leader's ability to transcend time also helps to explain why—as Weber emphasized—charismatic domination is a relatively unstable form of political order. Because charismatic time precludes regularized economic activity—which cannot take place without some sort of mundane synchronization of work schedules, whether according to natural cycles or the clock—there is strong pressure within social groups that accept a charismatic conception of time for that conception to give way, eventually, to "routinized" time of either a modern or traditional type. Whether the transcendence of time characteristic of charismatic domination is routinized into concrete, cyclical or abstract, linear time, or instead simply disintegrates, is thus from a Weberian perspective the central question in the analysis of changing patterns of social time perception.

Changing Conceptions of Time in Western Europe

How would a Weberian approach to social time perception help us to understand the historical shift from a traditional to a rational conception of time in western Europe more fully than modernization theory, Marxism, or cultural relativism? To begin with, a Weberian approach would depart from the emphasis of both Marxism and modernization theory on technological and material changes accompanying this process to include a more detailed analysis of religious and philosophical developments that affected the worldview of institution-building elites in western Europe concerning issues of time. Although Weberian analysis by no means ignores the material interests accompanying changes in spiritual and ideal interests, it nonetheless postulates that decisive transformations in social structures are not brought about by changes in material conditions alone. Rather, Weber argued, existing routines of social interaction generally are permanently altered only through the impact of new belief systems that are perceived as charismatic by significant numbers of people.

In explaining the historical pattern of time perception and use in western Europe, then, we must first examine the more historically common routinization of charismatic time into traditional time, in order to provide a comparative reference for evaluating what is unique in the historical development of time perception in the West. We must then focus our attention on the major idea systems that lie at the foundation of Western society—Judaism, Catholicism, and Protestantism—in order to evaluate the traditional, modern, and charismatic elements of each in their attitudes toward time. Finally we must turn our attention to the question of how, in the case of Protestantism, an originally charismatic force led to the routinization of time perception and use in a modern direction.

Let us first examine the process by which charismatic time routinizes in a traditional direction. It must be noted in this context that the traditional conception of time, unlike the rational time grid, still leaves substantial room for charismatic time transcendence within the social order. More precisely, while traditional cultures organize day-to-day economic activity according to concrete cycles based on seasonal and cosmic changes, traditional elites often claim special access to another order of reality lying outside time altogether—not the profane world subject to temporal constraints but the sacred realm of eternity. As Eliade has insightfully argued, traditional religions allow for "two kinds of time—profane and sacred. The one is an evanescent duration, the other a 'succession of eternities,' periodically recoverable during the festivals that make up the sacred calendar."[39] The drudgery of a life subject to mundane temporal cycles is thus periodically infused with spirituality and meaning on sacred festival days, when ordinary time is suspended. Traditional political legitimacy, therefore, depends on the claim of the leader or leaders to be able to guarantee this periodic infusing of the cycles of temporal existence with some element of sacred time transcendence—by establishing a direct line of succession from a superhuman founder or by demonstrating some sort of renewable divine mandate for rule.

When the connection of cyclical time with the sacred realm breaks down in a traditional setting—because of the disruption caused by war, famine, or natural disasters, for example—the very structure of time seems to loosen, and the stage is set for charismatic leaders of the pure type to claim authority. As Weber puts it, "the natural leaders in times of distress . . . were neither appointed officeholders nor 'professionals' in the present-day sense . . . but rather the bearers of specific gifts of body and mind that were considered 'supernatural' (in the sense that not everybody could have access to them)."[40] From the perspective adopted here, we might say that the charismatic leader in a traditional setting claims to provide *direct* access to the time-transcendent sacred realm that was previously mediated through the cycles of succession of the traditional chieftain, ruler, or patriarch. At the same time, successful charismatic leadership itself tends to generate new, concrete lines of succession from the leader, who may then be perceived as a semidivine founder. Economic activity, once the crisis has passed, resumes its cyclical and leisurely character, and the welding together of traditional, concrete time with sacred time is reaccomplished. In order to explain the rise of rational time as a dominant social organizing principle, then, we must first explain the establishment of the belief that disciplining one's activity according to an alternative abstract, linear understanding of time could uniquely guarantee access to the sacred realm of time-lessness—seemingly quite a paradox!

The origins of the rational view of time can be traced ultimately to Judaism.[41] By resting its very claim to authority on the basis of a non-repeating history, Judaism both undermined the cyclical metaphor for time and proclaimed that an alternative, linear temporal understanding could become the basis for the superiority of the group that adopted it. However, Judaism by no means accepted the notion that time is an abstract grid outside events; indeed, it is in the very concreteness of the genealogical links between Moses and the prophets and present-day Jews that the historical claim to unique status is based. Judaism still organizes social activity in a fundamentally traditional way, with six days of mundane activity giving way to a sacred Sabbath every week. But the emphasis Judaism places upon a linear progression outlined by historical Scriptures in establishing the claim to sacredness contained an element of abstraction in time orientation that could potentially be expanded into new realms.

This element was further reinforced by Christianity, which provided for the appearance of God himself at a particular time in human history. Because of the importance of this singular event, early Christians, especially Saint Augustine, heaped scorn upon pagan understandings of history as a series of repeating cycles.[42] Like Judaism, however, Catholic Christianity did not displace the prevailing conception of time in everyday social life as the concrete and cyclical passage of events. Catholicism continued to rest on the idea that time and events are intrinsically connected in the realm of the profane, whereas outside time *and* events altogether lies the sacred realm of God—a realm that could be made accessible to common humanity only by the leaders of the church hierarchy through the ceremony of the Eucharist. Thus, despite their linear understanding of history, neither Judaism nor Catholic Christianity provided any incentive to organize the time patterns of daily existence in any new way, since daily existence for both is still seen as profane and therefore subordinate to the charismatic, sacred realm beyond time.

During the Protestant Reformation, however, the Catholic separation of sacred time transcendence from profane social time came under theological attack. In attempting to break the Catholic Church's monopoly on the sacred realm, Protestant reformers such as Luther and Calvin argued that the most important expression of one's faith in God lay in the performance of everyday life tasks.[43] At the same time, the concrete link between Saint Paul and the present-day popes was denied and the ideal of early Christianity held up as a model to be emulated. Taken together, these two ideas tended to undermine the possibility of attaining access to the sacred realm through participation in present-day church rituals.

Protestantism thus undermined the remaining elements of traditional, concrete time in Catholic Christianity.[44] Henceforth Protestant believers were to conceive of themselves as individuals struggling to find a way to express their devotion to God *within* profane time. Time itself now appeared increasingly abstract, stretching out indefinitely into past and future, with more than a millennium and a half interposed between the believer and Christ's appearance on earth. In this context, the nature of the sacred realm itself had to be radically reconceived. For Protestants, no mere priest or pope could suspend the unfolding of a historical time line that encompassed absolutely everything that had ever existed; only the Second Coming of Christ could accomplish this. But how could individuals find access to the sacred in the meantime, now that the church could no longer guarantee it? The mushrooming of various Protestant sects after Luther's break with Rome can be understood as reflecting the struggle to find an authoritative answer to this question.

In this context, as Weber has argued, Calvinism played a crucial role, for Calvin's theology represented the most consistent attempt to work out a version of Christianity that would be compatible with a conception of time as an abstract line outside any particular events. Like Luther before him, Calvin placed great emphasis on the importance of an individual's work in a "calling" in sanctifying time-bound activity and in demonstrating Christian faith in a world without priests. But where Luther had defined the calling in traditional terms—as a way of life inherited from one's family and community—Calvin conceived of work in the calling as taking place in the context of a wholly individualized, abstract time. Every minute, every second represented an opportunity to serve God through productive activity—or reject God by wasting time in profane pursuits.

Thus, Calvinism's acceptance of an abstract and linear conception of time leads to a very different conception of the interrelationship between God, humanity, and time than all previous charismatic forces, for once time is seen as an abstract line encompassing all events, there can be nothing outside time—which means, paradoxically, that *time is the one thing that exists timelessly*. Since the sacred realm in traditional Christianity (and in other decisively traditional religions as well) had always been coterminous with the realm of the eternal, the effect of Calvin's theology was to identify charisma with the unfolding of time itself. For Calvin, then, linear time itself now belonged to the sacred realm, along with God—that is, linear time in Calvinism *becomes charismatic*. Acting in a way that shows an acceptance of the idea that time is holy is, for Calvinism, equivalent to acting in a way that accepts God as holy. The believer is therefore compelled to organize all mundane activity according to abstract time constraints or be cut off from the sacred realm altogether. From this it is but a short step to the secularization of abstract, linear time in Western culture summed up in Benjamin Franklin's formula "Time is money."[45]

Ultimately, this religious doctrine inspired elites in culturally Protestant countries to create a fundamentally new type of social organization in which, for the first time, abstract time discipline became more important for economic activity than concrete cosmic and seasonal cycles. During the industrial revolution, this new conception of time became the basis for the way of life of millions of people in northwestern Europe, chiefly through the widespread use of draconian time discipline in factory production. The traditional work patterns of peasants, with their reliance on imprecise time measures such as the daily passage of the sun through the sky and its seasonal fluctuations, were violently confronted with a new time norm based on "punching the clock," under which even a few minutes' lateness was cause for a severe cut in pay, if not dismissal.[46]

Instituting this "rational" time discipline was an intensely contested and politically volatile process.[47] As recently as the late 1800s, train schedules in England could not be completely coordinated because small towns refused to set their church clocks according to Greenwich mean time. As one railway company director stated, in terms more prescient than he realized, "I believe that [standardizing timetables] would tend to make punctuality a sort of obligation."[48]

By the beginning of the twentieth century, though, such pockets of resistance to a uniform rationalization of time along modern lines in liberal capitalist countries had been all but eliminated. Cyclical, concrete time was preserved only in a few realms, notably child rearing (which was considered as an exclusively female domain), sports (where cricket and baseball maintained the traditional attitude that the time of a game should be determined by how long the events of the game themselves last), and academia (which preserved a cyclical semester system and a traditional conception of leisure institutionalized as tenure).[49] Recently, even these realms have found themselves increasingly subject to rational norms of time organization: the semester system and tenure in academia have come under attack as being inefficient; new professional sports, such as American football and basketball, have become increasingly subject to the clock; and the raising of children in the household is being replaced (especially among the elite) by professional child care during the working hours.[50] Thus, a conception of time that had its roots in Judaism and Catholicism, and was then formalized in Protestantism and particularly Calvinism, has proven powerful enough—and, for a crucial segment of the population, inspiring enough—to become the dominant organizing principle of practically every realm of life in capitalist society.

The success of rational time as the basis for the enormously powerful political and economic institutions of liberal capitalism did not mean that alternative understandings of the charismatic realm of sacred timelessness disappeared entirely. A wide range of important sociological phenomena in modern societies attest to this, ranging from the continuing ability of prophets proclaiming the end of time to gain the allegiance of significant numbers of people, to the quite widespread experience of romantic love as a time-transcendent force that overwhelms one's ability to pursue one's mundane material interest.

However, within the cultural context of rational time, attempts to preserve some conception of a transcendent sacred realm have faced a distinctive problem: namely, that the idea of time as an abstract grid outside all events makes it extremely difficult for charismatic prophets or heroes to mark out a distinctive sacred history, different in kind from ordinary, mundane history. No matter how glorious one's victory in battle, no matter how remarkable one's ability to foresee the future, in the realm of abstract, linear time one's actions are still inescapably tied to their date of origin and hence comparable to other ordinary, time-bound phenomena. Ultimately, the only way to defend the sacred realm in a world subordinated to the rational time grid is to reconceive of linear history itself as infused with some kind of higher charismatic essence. As we shall see, the German philosophical tradition that gave rise to Marxism was centered on precisely this problem.

Charismatic-Rational Time in Leninist Regimes

To apply the Weberian framework for the study of social time outlined above to the case of Soviet Leninism is the task of the remainder of this work.[51] I endeavor to show that both the design of Leninism's core institutions and the patterns of political struggle among communists from 1848 to 1991 were rooted in the conception of time set out in the work of Karl Marx. Ken Jowitt is essentially correct to see Leninism as based on a novel form of institutionalized "charismatic impersonalism"—but the origins of this regime type can ultimately be traced to Marx's articulation of a conception of time that was itself an amalgam of charismatic and rational elements.[52] Prior to the establishment of the Soviet state, charismatic domination had everywhere opposed itself to any economic regulation according to abstract temporal units. Marxism, in this context, was based upon a fundamentally novel belief: that in the modern age, only those who learned to master abstract, rational time might succeed in gaining ultimate control over it.

In its promise to overcome ordinary time, Leninism was like other types of charismatic domination. But because of its embrace of rational time management as a means to this end, Leninist elites were able to do something Weber himself never imagined possible: create essentially charismatic state bureaucracies and, ultimately, a charismatic socioeconomic system. The synthesis of charisma and rational-legal proceduralism characteristic of both the Leninist polity and the Stalinist economy thus invalidates analytic dichotomies between "Utopia" and "development" in Soviet history.[53] Instead, both the political institutions established after the Bolshevik revolution and the economic institutions set up during the First Five-Year Plan were built upon the principle that "Utopia" could be made feasible by building time transcendence into time-bound development itself.[54]

This study begins by analyzing the *theoretical cycle* of development in the history of charismatic-rational time, from the writing of the *Communist Manifesto* to the breakup of international socialism during World War I. The Bolshevik revolution of 1917 is then interpreted as the beginning of the *political cycle* of development, which ends with the victory of Stalin over his rivals in 1928. The launching of the First Five-Year Plan marks the beginning of the *socioeconomic cycle*; this cycle ends with the decline of the planning system under Brezhnev. Finally, Gorbachev's *perestroika* is interpreted as an attempt to initiate a final, cultural stage of development in line with the previous stages: that is, the development of an entire society based upon revolutionary time discipline.

The charismatic-rational conception of time, like all forms of charisma, was subject to the process of routinization. But the complex mixture of time-denying and time-embracing impulses contained in Marxism, Leninism, and Stalinism was prone to decay in a rather complex manner. Like an elementary particle that can exist only at artificially induced high speeds, the synthesis of charismatic time transcendence and rational time calculation mandated by communist theory tended to break down into more stable components during periods of stagnation in the revolutionary movement—in particular, after the death of the "revolutionary innovator" primarily responsible for initiating each cycle.

As we shall see, this process occurred in a parallel manner during each of the three stages of development outlined above. In each case, "right" or "revisionist" communists began to emphasize rational time and to downplay or reject altogether the idea that the revolutionary movement might overcome time itself. "Left" or "voluntarist" communists began to call for an immediate war against rational time constraints and to reject any need for the observance of rational time discipline by revolutionary activists. Finally, "center" or "orthodox" communists began to rest their claim to leadership within the movement on their fidelity to the principles of the "founders"—Marx, Engels, Lenin, and ultimately Stalin—thus in effect turning socialism into a sort of traditional religion. Charismatic-rational time, in essence, routinized in three directions at once: the "right" in a rational direction, the "left" in a more purely charismatic direction (which was itself then subject to further routinization or decay), and the "center" in a "neotraditional" direction.[55]

The initial emergence of the Marxist charismatic-rational conception of time can only be understood as an ideological reaction to the rapid spread of rational time organization throughout western Europe under the auspices of liberal capitalism. The next chapter therefore examines the philosophical works most critical for understanding how the problem of time was discussed in Marx's intellectual milieu—the writings of Kant and Hegel.

2

TIME
IN THE
WORKS OF
KANT AND
HEGEL

he gradual destruction of the traditional conception of time in western Europe and its replacement by the rational idea of time as abstract and linear was not a neutral process, politically, culturally, or intellectually. Indeed, the triumph of rational time gave rise to a whole set of unprecedented problems for men and women trying to understand their place in the universe, problems that demanded some sort of philosophical resolution, for such a time conception fundamentally relativizes all human experience, and indeed even the existence of human beings themselves. The traditional conception of time as a concrete cycle, though it could not guarantee permanent social stability, did allow for the comforting view that human beings and their social arrangements were somehow essential to the very nature of things in time. In addition, since traditional politics defended a conception of sacred timelessness lying outside the mundane realm, philosophical ideas of beauty, truth, and morality could be seen as themselves timeless in nature, unaffected by earthly natural or social changes.[1]

The new conception of time as linear and abstract, however, undermined one by one all such traditional understandings of humanity's relationship to the universe. First, during the Enlightenment, time-honored social practices were exposed as mere superstition; only rules based on pure, timeless reason could be seen as valid. The concept of "absolute time" proposed by Newton in 1687, which codified the abstract time-space grid as the foundation of modern science, appeared to demonstrate the conformity of the most fundamental laws of the universe with rational time. By the mid-nineteenth century, with the publication of Darwin's theory of evolution, even the human species itself was shown to have merely a contingent existence, beginning and presumably ceasing to exist at definite points on the historical "time line." Did not such a view of time make a mockery of all claims on behalf of particular social orders to timeless validity? And how could human ideas of "morality" and "truth" ever be seen as more than temporary human constructs based on mere social contingencies?

Philosophers in western Europe from the Enlightenment onward naturally wrestled with precisely these sorts of questions. In England and Scotland, where Protestantism and the industrial revolution had decisively triumphed at an early date, philosophers such as Adam Smith and, later, the utilitarians began to elevate the satisfaction of purely temporal, empirical human interests into a new standard of morality perfectly compatible with modern scientific reasoning.[2] But in Germany, split between Protestantism and Catholicism, at the margins of the social and philosophical revolution of modernity in northwestern Europe, thinkers still sought to reconcile the emerging order with older notions of timeless morality and truth. The traditional social order had allowed for charismatic understandings of goodness, justice, and knowledge as eternal and sacred; the philosophers of the German Enlightenment—above all Kant and Hegel—were determined to find a way to give these timeless concepts meaning within the rational temporal framework as well.

This chapter shows how the philosophical resolution of problems of rational, linear time in German intellectual history gave rise to a fundamentally novel conception of time, one that saw time as an all-embracing force that was nonetheless essentially subordinated to a higher charismatic agent. It is this "charismatic-rational" time conception that was eventually inherited and ideologically transformed by Karl Marx. Moreover, future patterns of political differentiation that would emerge in fully developed form in Soviet history—in particular, the definitions of *left*, *right*, and *center* employed in Bolshevik factional struggles—were already foreshadowed in the philosophical debates immediately following the death of the originator of this time conception, G. W. F. Hegel.

Immanuel Kant and the Bifurcation of
Charismatic and Rational Time

The philosophy of Immanuel Kant sought to provide a systematic, universally valid codification of the forms and boundaries of knowledge. Whatever his success in this larger project, Kant's philosophy does provide an elegant framework outlining the assumptions of modern epistemology and ontology. Kant's three great critiques of the realms of reason—*The Critique of Pure Reason*, setting out what Kant sees as the necessary assumptions for the scientific understanding of empirical reality; *The Critique of Practical Reason*, which does the same for moral understanding; and *The Critique of Judgment*, which deals with the realm of taste—provided a division of philosophy into moral, scientific, and aesthetic spheres that has continued to exercise a powerful hold on the modern mind.

At the same time, Kant's philosophical system shows the a priori conditions of modern epistemology and ontology so clearly precisely because it stands to some extent *opposed to* the modern worldview. His *Critiques* are indeed critiques, designed to elucidate the *limits* of modern scientific reasoning, moral reasoning, and aesthetic reasoning. Put differently, Kant is able to provide a codification of modern reason only by observing it from a critical distance. As shown below, this critical distance is provided by his refusal to accept the rational conception of time as absolute in all spheres of understanding. This sets him apart from previous Enlightenment thinkers and explains his lifelong disputes with the British utilitarians, for whom the absolute nature of abstract, linear time was unquestioned.

Concerning the nature of time as it relates to scientific understanding, Kant has no dispute with the modern temporal outlook as established by Newton.[3] In the *Critique of Pure Reason*, Kant makes it quite clear that abstract time (as well as space) must simply be accepted, a priori, for the rational ordering of our sense perception of empirical reality to be possible: "Time is a necessary representation that underlines all intuitions. We cannot, in respect of appearances in general, remove time itself, though we can quite well think of time as void of appearances. Time is, therefore, given *a priori*. In it alone is actuality of appearances possible at all. Appearances may, one and all, vanish; but time (as the universal condition of their possibility) cannot itself be removed."[4] In addition, Kant argues that our a priori conception of time must not only be abstract, since time can be thought of as "void of appearances," but also linear—time does not fold in on itself: "Time has only one dimension; different times are not simultaneous but successive (just as different spaces are not successive but simultaneous)."[5] In all this, Kant's view of time is ideal-typically modern according to the typology developed in the preceding chapter.

However, in Kant's philosophical system as a whole, abstract rational time does not govern every realm of understanding. Even where Kant sees abstract time as necessary to human understanding—that is, in the sphere of pure reason—time nonetheless remains an a priori metaphysical supposition, not an empirical fact. When Kant turns his attention to moral philosophy, he stresses the need to abandon time-bound reasoning altogether. For moral reasoning to be possible as a sphere separate from scientific, empirical understanding, Kant argues, we must avoid tying moral arguments either to past constraints on or to the future consequences of our behavior: "Not only are moral laws together with their principles essentially different from every kind of practical cognition in which there is anything empirical, but all moral philosophy rests entirely on its pure part. When applied to man, it does not in the least borrow from acquaintance with him (anthropology) but gives *a priori* laws to him as a rational being."[6]

Theories of morality that *are* based on such a "prior acquaintance" with man—such as traditionalism, which asserts that morality lies in the upholding of empirically existing customs, or utilitarianism, which argues that morality must be based on a calculation of empirically given interests—must either deny that morality has any rational basis or collapse moral reasoning into scientific reasoning, thus reducing "morality" to a mere word applied to essentially predetermined, time-bound behavior. Instead, Kant proposes that we derive universal moral laws, valid for any rational being (human or otherwise), by imagining ourselves in a timeless "kingdom of ends" in which we relate to each other solely as rational beings.[7]

Kant's moral philosophy becomes enmeshed in sticky difficulties when he attempts to derive concrete universal moral rules on this basis.[8] However, for present purposes, it is more important to stress the separate relationships to time that underlie the separate realms of scientific and moral understanding in Kant's philosophical system. Rational, linear time is held to be necessary a priori for pure reason; but an essentially charismatic, time-transcendent perspective is held to be the basis of all true moral reasoning. In order to see oneself both as empirically existing and as a moral agent, Kant seems to be saying, we must be schizophrenic, imagining ourselves to be simultaneously living in time-bound reality *and* remaining metaphysically in a timeless "kingdom of ends."

This tension, difficult to accept even in theory, becomes even more troubling once we begin to apply it politically. Specifically, the question arises: Is the political sphere to be considered an "empirical" reality, and thus bound to linear cause-and-effect determinism, or a "moral" reality, and thus subject to universal moral laws deduced a priori from a timeless conception of ourselves as rational moral agents? Kant's own answer to this question is decidedly ambiguous. In *Perpetual Peace*, written toward the end of Kant's life, he argues that any political reasoning worth the name must be seen as a subset of moral reasoning: "There can be no conflict between politics, as an applied branch of right, and morality, as a theoretical branch of right (i.e., between theory and practice); for such a conflict could occur only if morality were taken to mean a general doctrine of expediency, i.e. a theory of the maxims by which one might select the most useful means for furthering one's own advantage—and this would be tantamount to denying that morality exists."[9]

But if this is the case, how does political action link up with the scientific understanding of human beings not as pure moral agents but as creatures with empirical interests, bound to biological and historical laws? Here Kant hedges his bets, placing his faith in yet another a priori assumption—that empirical history is inherently morally progressive, despite our existing moral flaws, since to assume otherwise is to make a mockery of the rationality of God's creation and our role in it. As Kant argues, the idea of historical progress

> opens up the comforting prospect of a future in which we are shown from afar how the human race eventually works its way upward to a situation in which all the germs implanted by nature can be developed fully, and in which man's destiny can be fulfilled here on earth. Such a *justification* of nature—or perhaps of *providence*—is no mean motive for adopting a particular point of view in considering the world. For what is the use of lauding and holding up for contemplation the glory and wisdom of creation in the non-rational sphere of nature, if the history of mankind, the very part of this great display of supreme wisdom which contains the purpose of all the rest, is to remain a constant reproach to everything else? Such a spectacle would force us to turn away in revulsion, and, by making us despair of ever finding any completed rational aim behind it, would reduce us to hoping for it only in some other world.[10]

This argument, however—which is in any case rather sketchily laid out in the context of Kant's work—is clearly unsatisfactory. Even an acceptance of Kant's somewhat tentative faith in historical progress does not in itself provide any reason for preferring one concrete action to another. Such a guide to action could be found only in a philosophy of history that explicitly subordinated the empirical content of history to an a priori scheme of human progress—and this Kant is understandably hesitant to do: "It would be a misinterpretation of my intention to contend that I meant this idea of a universal history, which to some extent follows an *a priori* rule, to supersede the task of history proper, that of *empirical* composition."[11] Whatever the connection between moral reasoning and empirical interests may be, they cannot be determined specifically once one has accepted the absolute division of moral reason and empirical reason at the core of Kant's philosophy. In the end, Kant leaves human actors suspended between the empirical realm, governed by abstract and linear time, and the moral realm of time transcendence, inviting us simply to live with the tension.

Hegel's Synthesis of Charismatic and Rational Time

It is precisely Kant's absolute division of moral and empirical reason that Hegel attacks in his attempt to reconcile eternal meaning and the time-bound processes of history.[12] Hegel, like Kant, attempts to escape the seeming relativism of all political and moral principles from a rational time perspective by seeing the process of history as a drama of progress toward greater and greater freedom. However, unlike Kant, Hegel does not postulate this idea of progress as a mere a priori principle that allows us to make moral sense of history; rather, he understands the progressive nature of history as an absolutely fundamental element of empirical history itself. He therefore does not worry that his idea of history as progressive will "supersede the task of history proper," as Kant feared; for Hegel, the empirical and metaphysical aspects of history, properly understood, *necessarily coincide.*

Thus, Hegel begins his lectures on the philosophy of history by asserting that "the sole thought which philosophy brings to the treatment of history is the simple concept of *reason*: that reason is the law of the world and that, therefore, in world history, things have come about rationally."[13] Yet soon afterward he insists that "history itself must be taken as it is; we have to proceed historically, empirically," avoiding "a priori historical fiction."[14] On Hegel's view, these two statements are not contradictory; there simply can be no conflict between the empirical substance of history and its formal meaning; moral and empirical reason are seen to be intrinsically linked rather than necessarily separate. Hegel thus collapses the empirical realm and the moral realm, which for Kant were governed by different types of time. The question therefore arises: How does Hegel reconcile the charismatic, timeless quality of freedom as the goal of history with the seemingly modern view of time behind his abstract conception of history? What is Hegel's basic view of time?

Hegel's belief in progress would seem at first glance to indicate an abstract and linear time perspective. His desire to overcome purely traditional, cyclical time, at least, is quite clear: "Historical change, seen abstractly, has long been understood generally as involving a progress towards the better, the more perfect. Change in nature, no matter how infinitely varied it is, shows only a cycle of constant repetition. In nature nothing new happens under the sun, and in this respect the multiform play of her products leads to boredom."[15] Hegel goes on to label peoples who see their lives as subject to cyclical time constraints as "pre-historical"—a term that is obviously quite pejorative in the context of Hegel's philosophy of history.[16]

On closer examination, however, Hegel's view of time is substantially more complex than a simple rejection of traditional cycles for modern linearity. Indeed, in the *Phenomenology of Spirit*, Hegel explicitly rejects the Newtonian view of time as a mathematically divisible, abstract dimension complementing the three dimensions of space, arguing that this definition reduces time to a "paralyzed form" that "cannot cope with [the] sheer unrest of life."[17] Instead, from a philosophical perspective, time must be seen as encompassing the transitoriness of all concretely existing things; nonetheless time itself, seen abstractly, remains a sort of stasis that is ultimately unaffected by the birth and death of particular empirical phenomena: "Appearance is the arising and passing away that does not itself arise and pass away, but is 'in itself' [i.e., subsists intrinsically], and constitutes the actuality and the movement of the life of truth. The True is thus the Bacchanalian revel in which no member is not drunk; yet because each member collapses as soon as he drops out, the revel is just as much transparent and simple repose."[18] For Hegel, then, it is *not* the passage of time per se that brings forth progress in history. Indeed, from the perspective of finite beings, time represents a destructive force causing the death and decay of all present existence.

In Hegel's view, then, for progress to occur despite the constant destructive action of time, some powerful agent must be operating in history, overcoming abstract, rational time's "negativity." Hegel names this agent "Spirit" (*Geist*); he asserts that "world history in general is the development of Spirit in *time*."[19] Hegel's view of time, therefore, despite elements of modernity in his attempt to be faithful to scientific and empirical method in his charting of historical development, must be seen as essentially charismatic, based on the idea that time can be overcome by a more powerful agent, Spirit. As Hegel puts it, "only the changes in the realm of Spirit create the novel."[20] In his introduction to the *Philosophy of History*, Hegel argues that the overcoming of time is the symbolic first step beginning the historical march of Spirit: "At first [Chronos] ruled, time itself—the golden age without moral works. What it produced, its children, were devoured by it. Only Zeus, who gave birth to Athena out of his head . . . conquered time and set a limit to its lapse. He is the political god, who has produced a moral work, the state."[21] The course of history from that point onward must be seen as the process by which Spirit overcomes all the particular contingent temporal destructions of individuals and of nations, until its final achievement of complete self-consciousness: "The Spirit, devouring its worldly envelope, not only passes into another envelope, not only arises rejuvenated from the ashes of its embodiment, but it emerges from them exalted, transfigured, a purer Spirit. It is true that it acts against itself, devours its own existence. But in so doing it elaborates upon this existence; its embodiment becomes material for its work to elevate itself to a new embodiment."[22] It is Spirit, then, that is eternal and essential, while time is only the framework in which the development of Spirit takes place.

The question of whether time is "really" cyclical and concrete or linear and abstract is thus, in Hegel's view, a misleading one. From the charismatic, time-transcendent perspective of Spirit, temporality itself is in the final analysis only a lengthy passage through an extended now:

> In dealing with the idea of Spirit only and in considering the whole of world history as nothing but its manifestation, we are dealing only with the *present*—however long the past may be which we survey. . . . This implies that the present stage of Spirit contains all previous stages within itself. These, to be sure, have unfolded themselves successively and separately, but Spirit still is what it has in itself always been. The differentiation of its stages is but the development of what it is in itself. The life of the ever-present Spirit is but a cycle of stages, which, on the one hand, co-exist side by side, but, on the other hand, seem to be past. The moments which Spirit seems to have left behind, it still possesses in the depth of its present.[23]

Here Hegel seems to suggest that the timeless perspective of Spirit reconciles the conflict between our limited conceptions of time as either linear or cyclical; the unfolding of Spirit in history takes the form of a "cycle of stages," but the essence of Spirit itself remains unchanged.

However, as in Kant's philosophy, Hegel's attempt to overcome the conflict between the timeless quality of freedom and the temporal flux of empirical reality gave rise to difficulties concerning the problem of human action. If for Kant action was suspended precariously between the timeless moral realm and the time-bound empirical realm, for Hegel, who denies any conflict between eternal values and temporal processes, individual action appears to become simply irrelevant: if the march of Spirit is and must be manifest in historical development, the acts of mere mortals would seem to be powerless to influence future outcomes. The question becomes, how can human beings gain earthly access to the realm of Spirit that can alone provide their existence with transcendent meaning?

One potential solution to this problem is set out in Hegel's theory of the state. Unlike Kant, Hegel developed his political philosophy in great depth, chiefly in his *Philosophy of Right*.[24] In this work he tries to show how the development of the modern state is a concrete manifestation of the progress of freedom that is for Hegel synonymous with Spirit. From the early unconscious corporate solidarity of the family, he argues, human social life has progressed first toward the development of civil society, characterized by economic individualism and the abstract and formal freedom of citizenship. However, civil society by itself is an incomplete expression of freedom, because while it breaks down the solidarity of the family, it provides no alternative community that could become the focus of political membership and obligation, and thus its legal freedoms remain abstract, unconnected to any sense of duty to others in real political life. Such an "ethical life"—as opposed to the mere "morality" of civil society—can be provided only by the modern state.[25] The state does not destroy civil society or the family but instead reconciles the abstract freedom of the former with the sense of empirical communal membership of the latter, creating a higher free, collective entity.

However, this solution, despite its elegance and insight, creates new difficulties, for the establishment of human freedom here appears more as a macrohistorical process than as a goal toward which individuals can meaningfully contribute. Furthermore, different interpretations of Hegel's historical dialectic of freedom might lead to different prescriptions for political action. Is the empirical modern state—for Hegel, the Protestant monarchy of nineteenth-century Prussia—itself to be thought of as a temporal phenomenon, and therefore subject to the "negativity" of time's destructive power? If so, then our allegiance to it can only be conditional, as the march of Spirit in history will continue to unfold in new political forms that represent even higher expressions of freedom. But if not—if the empirically existing state in which we happen to live is to be thought of as beyond the corrosive power of time—then the state itself becomes the absolute and final manifestation of Spirit on earth, and in opposing it we necessarily oppose freedom and meaning in history. Both interpretations of Hegel's meaning gained currency even during his lifetime, and the battle between them continues unabated in the literature.[26]

Hegel, I would argue, ultimately avoided any clear resolution of this problem by taking up an essentially passive philosophical perspective. At several points in his work, Hegel adopts the tragic mode in his depictions of historical process; in these passages human action seems either only accidentally efficacious or wholly impotent. Hegel does allow for "world-historical individuals . . . who grasp . . . a higher universal, make it their own purpose, and realize this purpose in accordance with the higher law of the Spirit."[27] However, Hegel insists that "such individuals have no consciousness of the idea as such. They are practical and political men"; they grasp the Spirit of their age only intuitively and thus cannot be fully aware of the consequences of their actions.[28]

For those individuals who are not "world-historical," the consequences of Hegel's tragic view of human action are more stark. Hegel states that "the universal law is not designed for individuals, as such, who indeed may find themselves very much the losers"; even world-historical individuals themselves "must trample down many an innocent flower, crush to pieces many things in [their] path."[29] Such a tragic view of the march of Spirit sees the rise and fall of all social institutions as part of a higher developmental history to which politics provides only accidental access. The ultimate futility, in Hegel's view, of trying to hold back the destructive force of time through political activity is clear from the following passage:

Spirit is essentially the result of its own activity. . . . We can compare it with the seed of a plant, which is both beginning and result of the plant's whole life. The powerlessness of life manifests itself precisely in this falling apart of beginning and end. Likewise in the lives of individuals and peoples. The life of a people brings a fruit to maturity, for its activity aims at actualizing its principle. But the fruit does not fall back into the womb of the people which has produced and matured it. On the contrary, it turns into a bitter drink for this people. The people cannot abandon it, for it has an unquenchable thirst for it. But imbibing the drink is the drinker's destruction, yet, at the same time the rise of a new principle.[30]

In this vision of the inevitable destruction of peoples who have an "unquenchable thirst" for continuing their once successful patterns of action, Hegel's essentially tragic stance is manifest.

Ultimately, Hegel upholds philosophical contemplation of history, not the making of history, as the highest activity, for only philosophy can consciously understand the workings of Spirit. Nor does Hegel hold out any possibility that philosophers themselves can become effective social actors; when it comes to advice as to what the world "ought to be," Hegel writes, "philosophy . . . always comes on the scene too late to give it. As the thought of the world, it appears only when actuality is already there cut and dried after its process of formation has been completed. . . . When philosophy paints its grey in grey, then has a shape of life grown old. By philosophy's grey in grey it cannot be rejuvenated but only understood. The owl of Minerva spreads its wings only with the falling of the dusk."[31] Such an understanding of the task of philosophy, it would appear, can only leave the realm of action in an eternal twilight.

However, on the intellectual level, at least, Hegel had formulated a stunningly original response to the conflict between "eternal" philosophical values and the modern "scientific" time grid. Proposing a charismatic agent, *Geist*, so powerful that it could not only triumph over the cycles of nature but even transfigure the abstract, linear time in which the laws of science operated, Hegel could argue that empirical, time-bound existence was necessarily always infused with transhistorical meaning. The historical destruction of old value systems taking place throughout Europe could thus be understood as necessary for the reconstruction of social values on a higher level of freedom and rationality. Despite the complexity and even obscurity of Hegel's style of argumentation, students working with him found this breakthrough inspiring, even life transforming. As one of them wrote in 1830, a year before Hegel's death: "You will think I am insane when I tell you I see God face to face, but it is true. The transcendent has become immanent, man himself is a point of light in the infinite light and like recognizes like. Because I myself am all essence I can know all essence, and insofar as I rest in the great heart of God I am blessed here and now. Oh, if I could only describe to you how blessed I am."[32] Hegel's achievement—which appears to have transformed Hegel himself into a sort of charismatic leader for his most admiring intellectual disciples—marks the birth of the "charismatic-rational" conception of time, whose institutional development in Marxist theory, Leninist politics, and Stalinist economics we will trace in the chapters to follow.

The Hegelian School and the Emergence of Left, Right, and Center

For German intellectuals of the early nineteenth century, the sorts of philosophical dilemmas we have examined above were not merely problems of abstract logic. They were simultaneously issues of central importance to political elites of the period, for whom the problem of re-legitimating monarchical rule in the wake of the French Revolution and Napoleonic Wars was paramount. As is well known, Hegel himself remained throughout his life a steadfast defender of the Protestant monarchy of the Prussian state after the legal and administrative reforms instituted in response to Napoleon's conquests. Yet the increasingly antiliberal political trends toward the end of Hegel's life made it difficult to continue to see Prussia as the highest achievement of historical freedom. As Toews has shown, as early as the 1820s the Hegelian school had begun to split into three tendencies: "accommodationists" who supported the Prussian state unconditionally, critical reformers who thought of themselves as a kind of "loyal opposition" that would push Prussia back toward more philosophically "rational" policies, and those who hoped to transcend the Prussian state altogether in order to begin a new stage of historical progress.[33]

During Hegel's lifetime, the radically different political implications of these interpretations of his philosophy were tempered by the common adherence of all three camps to his basic methodology and language. But by the mid-1830s, these internal tensions, combined with a marked decrease of support by Prussian state authorities of Hegelian scholarship, led to overt splits in the Hegelian movement. The immediate catalyst for this factional breakup was a dispute about the relationship of Hegel's dialectic to the Christian faith. Hegel had always portrayed himself as an orthodox Lutheran and had been more or less accepted as such during his lifetime. However, even if one assumes *Geist* is synonymous with "God" in Hegel's work—an assumption that is at best problematic—it is not at all clear how Hegel's theory of history can be reconciled with the Christian portrayal of Christ's first and second coming.[34]

Of course, the Christian theological roots of Hegel's thinking are quite clear in his notion of history as an unfolding of spiritual truth, culminating in universal redemption. Jesus himself could be presented as the ideal synthesis of *Geist* and empirical existence that Hegel held to represent the complete attainment of freedom—and indeed, Hegel made this connection. But philosophically, what was one to make of the fact that Christ's appearance on the earth took place almost two millennia before the "highest" realization of Spirit in political form in modern Prussia? If the dialectic of Spirit and matter proceeded through the negation of outmoded social forms, then wouldn't the negation of Christianity itself be required at some point? While "accommodationist" Hegelians proclaimed that Hegel's system merely presented a comprehensive and "rational" interpretation of the eternal truth of Christ, more critically minded Hegelians tended to claim instead that Christianity was but a primitive statement of the higher truth of Hegelian philosophy, presented in a "mythical" form more comprehensible to the "limited" popular consciousness of Jesus' time.

The publication of *The Life of Jesus* by the radical Hegelian D. F. Strauss in 1835, which argued for the latter position, sparked an intense controversy that forced everyone within the Hegelian school to choose sides. By 1841, Strauss himself was able to identify three different factions within the Hegelian movement on this question:

> To the question of whether and how far the Gospel history is contained as history in the idea of the unity of divine and human nature, there are three possible answers: namely that from this concept either the whole Gospel narrative or only a part of it or finally that neither the whole of it nor a part of it can be deduced as history from the idea. If these three answers or directions were each represented by a branch of the Hegelian school, then one could, following the traditional simile, call the first direction the right, as the one standing nearest to the old system, the third left and the second the center.[35]

Despite the rather offhand way in which Strauss presented these categories, the labels stuck—in a way that was ultimately highly consequential for political organization and mobilization in Marxist and Bolshevik circles.

The "traditional simile" referred to by Strauss is, of course, the practice of referring to political positions as they were arrayed in the National Assembly during the French Revolution, where the strongest supporters of the monarchy were seated in the extreme right wing of the hall and the Jacobins at the extreme left. These political views happened to correspond precisely with those held by Hegelians with the various attitudes toward Christianity discussed above: those Hegelians who thought "the whole Gospel narrative" was still valid were staunch supporters of the Prussian king; those Hegelians who thought the gospel had been transcended were radical democrats; those who tried to adopt a moderate perspective on this issue tended to be moderate reformers politically as well.

However, the reinterpretation of the political categories of "left," "right," and "center" in terms of Hegelian philosophy gave the terms additional connotations—the echoes of which can still be felt in political discourse in modern Western societies. To be politically "right" now meant not only to be in favor of the monarchy or traditional authority but also to be "reactionary," attempting to keep history from moving forward. To be "left" now meant not only to support radical democracy but to be historically "progressive," bent on pushing the dialectic forward to "higher" levels. Finally, the "center" position now indicated not only political moderation but also a basic indecisiveness about one's ultimate "world-historical" commitments.

At a deeper level, the underlying cause of this tripartite differentiation in the Hegelian school was the ambiguity of Hegel's conception of charismatic-rational time as applied to the analysis of concrete political regimes. Toews has given a concise explanation of how different stances on the relationship of eternal Reason and temporal history led to the three different tendencies within Hegelianism: "Within this framework of explanation the focus of division was the interpretation of Hegel's identification of Reason and reality. Was this identification completed, an ongoing process, or a future goal?"[36] The first of these stances, that of "right" Hegelianism, meant that existing historical time was already saturated with charismatic validation of realized Spirit. Whatever already existed, from this point of view, was sacred, validated by *Geist*. The second stance, that of the "center" Hegelians, meant that time-transcendent Spirit and time-bound existence continued to interpenetrate each other in existing institutions. From this point of view, the sole charismatic force capable of comprehending this interpenetration and allowing human beings to take part in its "world-historical" resolution was the "Hegelian system" itself. The third stance, that of the "left," radically desacralized all present social forms as a "negation" of Spirit and argued that therefore the status quo itself needed to be negated in order to create a new, free humanity. Accordingly, the left Hegelians after 1835 began searching avidly for some wholly new charismatic force that would usher in the dialectical transcendence of the nineteenth-century social order.

Left Hegelianism did not cohere as a political movement for very long. Expelled from the German academic establishment, by and large ignored by the public, chronically short on funds, and constantly attacking one another's philosophical principles, the "Young Hegelians" (as they were also known) maintained a semblance of organizational unity for no more than a decade or so. Indeed, this should not surprise us; from the Weberian perspective, the left Hegelians amounted to a loose affiliation of would-be charismatic leaders with no following—hardly a recipe for organizational success.

But in one sense the Young Hegelians were radically different from all previous charismatic sects. As a result of Hegel's reconceptualization of the nature of history and eternity, his "left" followers were forced to search for charismatic transcendence not outside time and space but within worldly time-space itself. On this figures as diverse as Bauer, Feuerbach, and Marx were all agreed: the new millennium would begin on earth, not in heaven. History itself would be begun anew, and human beings would directly appropriate the powers of creativity and transcendence that had previously been "projected" into an imaginary "sacred" realm. The question, obviously, was how to bring this revolutionary transformation about historically. Such was the starting point of Karl Marx's theoretical work.

3

THE

THEORETICAL

CYCLE

From Marx to the
Second International

The thought of Karl Marx represents the intellectual and historical link between Protestantism and Western economic theory on the one hand and Leninism and Soviet economic practice on the other. In order to elucidate the distinctive features of the Leninist view of time, therefore, it is necessary to examine the treatment of time in Marx's work.[1] Despite the intellectual logic of this starting point, discussions of Leninist politics—let alone of Stalinist economics—rarely deal with Marx's writings in any detail, and therefore it is worthwhile to specify precisely what such an examination can and cannot accomplish.

Most important, a careful reading of Marx can neither exonerate him from responsibility for interpretations of his teachings by later revolutionary actors nor prove the existence of an ironclad logic of totalitarianism within Marx's thought that inevitably culminated in Stalinist terror.[2] Such approaches to Marx are simply not consistent with the historical record. There are, on the one hand, clear elements of continuity between Marx's and Lenin's theoretical works indicating that they shared a common worldview. Denying this fact would make Lenin's role in the history of the Marxist movement incomprehensible. However, we need not conclude from this that Lenin merely realized the predetermined fate of the communist ideal as formulated by Marx. Indeed, a wide variety of non-Leninist interpretations of Marxism were possible, as this chapter demonstrates. The argument that Leninism flowed inevitably from Marxism forfeits the ability to interpret Marx's ambiguities on many crucial issues as precisely that—ambiguities.

Both these interpretations therefore miss the real historical link between Marx and his Soviet disciples: namely, that while the consistent core of Marx's theoretical vision welded Marxists together—however loosely—into an identifiable movement, the ambiguities contained in Marx's thought set the parameters for the rival interpretations of socialism that would later become the main content of Marxist political debate. Indeed, Lenin's interpretation of Marxism was politically successful precisely because it maintained fidelity to the core of Marx's theoretical vision, while resolving some of the key ambiguities in Marx's work about how to put that vision into political practice.

To demonstrate this, however, we need to be able to identify both the consistent core and the unresolved ambiguities within Marx's theorizing. I argue in this chapter that the core of Marx's theoretical achievement was to transform the Hegelian synthesis of charismatic and rational time orientations into a basis for social action through his distinctive interpretation of the concept of communism—a concept that united an abstract theory of class struggle in historical time with a call for revolutionary praxis to transcend that history. Marx's work remained ambiguous, however, in its characterization of the political, socioeconomic, and cultural strategies necessary for realizing the proletarian revolution. On each of these issues, Marx's work remains fundamentally dualistic—sometimes relying on a charismatic interpretation of the revolution in which time constraints disappear, sometimes arguing that the need to conform to time constraints continues even after the proletarian revolution itself.[3]

The chapter concludes with an examination of the problems confronted by Marxists of the Second International who attempted to put Marx's theory into political practice after the founder's death in 1883. As in Germany after Hegel's death, distinctive left, right, and center interpretations of Marxism emerged and competed against one another. However, Marx's reconstruction of the Hegelian dialectic on a materialist basis changed the nature of these categories in subtle but important ways.

Marx's Conception of Time

Marx's intellectual debt to Hegel, as has often been noted, is an enormous one.[4] Despite Marx's biting polemics against various Young Hegelians, his early involvement with the Hegelian school had a decisive impact on his intellectual development. As demonstrated below, one of Hegel's most consequential legacies to Marx and Marxism is the essentially charismatic Hegelian conception of time. Marx himself, unlike his philosophical predecessors, did not concern himself with the question "What is time?" directly. However, in his early philosophical writings on social history and the role of human beings in shaping it, the same themes and paradoxes we have examined in Hegel's work reappear, and an implicit Marxian resolution of the conflict between charismatic and rational time is worked out. Since an analysis of the nature of time and its use under modern economic organization becomes quite explicitly one of the key elements of Marx's mature critique of capitalism, investigating the influence of Hegel's conception of time on Marx's thought is crucial to an understanding of Marx's general outlook—and to an understanding of the later debates among Marxists attempting to put Marx's ideas into practice.

From the outset, Marx shared the Kantian and Hegelian ambivalence about the conflict between modern scientific, empirical reasoning and seemingly "timeless" values such as freedom, personal fulfillment, and community that alone might impart meaning to human history.[5] From Hegel, he inherited a substantial part of his solution to that conflict: history was to be seen as a rationally understandable, empirical chain of cause and effect that nonetheless would lead dialectically to a universal human overcoming of oppression and social fragmentation. For Marx, as for Hegel, the historical path toward fulfillment necessarily involved a movement from primal, unconscious familial types of communal solidarity, toward the individualism and atomism of civil society, to a final overcoming of the opposition between individual and community in a fully free social order.[6]

However, despite the substantial congruence between the two thinkers, Marx fundamentally rejected Hegel's outlook on human action in both its statist and tragic aspects. As Paul Thomas has shown, in rejecting the supposed "universality" of Hegel's state, Marx instead argued "that modern political institutions were so organized as to make the state the mere instrument of the interests dominating civil society and masquerading as general 'universal' interests. The state to Hegel was a different axis; to Marx it merely provided one more set of conduits."[7] Nor could Marx accept any resort to metaphysical agents such as "Spirit" in accounting for meaning in history; such concepts denied any real control by human beings over their destiny and implied that calls for political action should be avoided by mature philosophers—a conclusion completely contrary to the instincts of the radical Marx. As he put it famously in his eleventh thesis on Feuerbach, "the philosophers have only *interpreted* the world in various ways; the point is to *change* it."[8] This desire to reunite theory and practice became the driving principle behind Marx's work.[9]

As we have seen, however, the logic of Hegel's philosophical system made it hard to justify human attempts to change the course of history: the entire weight of time, Hegel had argued, worked to thwart human action and to undermine the permanence of human institutions. Only something greater and more powerful than abstract time's negativity could rescue human existence in time from meaninglessness. Accepting Hegel's way of posing this problem, but rejecting any idea of *Geist* working in history, Marx had to find an empirical, human activity capable by itself of *mastering time within the very confines of time*. His early philosophical work argues that there is indeed such a force, both empirical and time transcending in nature: human labor.

Marx's *Economic and Philosophical Manuscripts of 1844* provide his most detailed philosophical examination of the role of labor in expressing human autonomy. There, Marx defines labor as "*life-activity, productive life* itself"; he goes on to say that "the whole character of a species—its species-character—is contained in the character of its life activity; and free, conscious activity is man's species-character."[10] Free, conscious labor is also what distinguishes humans from animals, since "the animal is immediately one with its life-activity," whereas "man makes his life activity itself the object of his will and of his consciousness."[11] In other words, while animals simply exist, unconsciously subject to temporal laws, humans possess the unique ability to produce through labor a future product that is a tangible manifestation of present will. Thus, for Marx, labor possesses on a mundane level the same ability to overcome time that Hegel attributed to Spirit alone.

Marx's grounding of human will and creative power in labor allows him to examine economic history in a new light, for if labor is the source of man's "species-character," then the study of the particular mode of production in which human labor is embedded in each society is necessarily of profound importance. As Marx and Engels explained in *The*

German Ideology: "This mode of production must not be considered simply as being the reproduction of the physical existence of the individuals. Rather it is a definite form of activity of these individuals, a definite form of expressing their life, a definite *mode of life* on their part. As individuals express their life, so they are. . . . Hence what individuals are depends on the material conditions of their production."[12] The history of the transition from earlier to later modes of production—from primitive communism, to slavery, to feudalism, to capitalism, and finally to full communism—is thus for Marx no longer a tale of the progress of Spirit but instead a chronicle of empirical human economic action shaping human history: "For the socialist man the *entire so-called history of the world* is nothing but the creation of man through human labor."[13] The idea that human labor is simultaneously creative and grounded in necessity allows Marx to imagine a practical path for the overcoming of time-boundedness: the emancipation of labor and the creation of a communist society in which people would be free to express their "species-character" in daily life activity.

But how is labor to be emancipated within the framework of time-bound, class society? Paradoxically, although Marx sees labor as the fundamental creative force in history, he does not see labor in itself as constituting a means for human emancipation. Labor in class society, Marx argues, is fundamentally "alienated," and maximally so in capitalist society. Alienated labor, unlike free, conscious labor, is *"external* to the worker, i.e., it does not belong to his intrinsic nature; . . . in his work, therefore, he does not affirm himself but denies himself, does not feel content but unhappy, does not develop freely his physical and mental energy but mortifies his body and ruins his mind."[14] If labor under current empirical conditions—caught within the abstract time line fundamental to scientific understanding—can only "mortify" the worker, Marx would seem to remain caught within the paradox of the Hegelian conception of time, needing some supratemporal charismatic force to rescue human beings trapped in empirical history from meaninglessness and heteronomy.

Marx cannot simply ignore one or the other of these two sides of labor in class society—its time-transcendent creative potential and its time-bound drudgery in actuality—since each provides an essential element of his attempt to reconcile human empirical existence in time and the possibility of overcoming time's dominion over humanity. On the one hand, Marx wants to insist that people, here and now, have an autonomous role in shaping history; this leads him to retain a charismatic role for human action, and more specifically for labor. On the other hand, Marx inherits from Hegel a keen sense for the overwhelming power of the temporal forces he is trying to overcome, arguing that social development is subject to historical forces that must necessarily affect labor as well. To deny the efficacy of human action in time would be to destroy any possibility of finding transcendent meaning in daily life, but to ignore the time-boundedness of human action would be to become a foolish Utopian. How can these seemingly incompatible faiths—in the power of historical conditioning *and* in the potential for historical transcendence of that conditioning by human agents—be reconciled?

Marx's answer to this question is unambiguous: only through revolutionary action can the creative potential of human labor within the context of time-bound socioeconomic laws be fully actualized. It is not labor per se but collective revolutionary action by laborers that is uniquely capable of collapsing time within time. As Marx and Engels state in one highly revealing passage:

> Both for the production on a mass scale of this communist consciousness, and for the success of the cause itself, the alteration of men on a mass scale is necessary, an alteration which can only take place in a practical movement, a *revolution*; this revolution is necessary, therefore, not only because the *ruling* class cannot be overthrown in any other way, but also because the class *overthrowing* it can only in a revolution succeed in ridding itself of all the muck of ages and become fitted to found society anew.[15]

Here the authors hit at the very heart of the dilemma. For human beings to act effectively within a temporal history governed by inexorable economic laws, they must be "altered on a mass scale," made somehow capable of transcending those very laws. Therefore, only a "practical" movement for such transcendence, a revolution, can clear away the entire weight of time—the "muck of ages"—to allow for a new society based on the absence of previous historical constraints on human development. Proletarian revolutionary action, then, ends up playing the role of the charismatic force capable of conquering time that labor alone cannot.

It is in this sense that Marx redefines the concept of communism to mean, essentially, the social movement uniting all proletarian revolutionary actors—the movement within time that liberates humanity from time. The term *communism* had been introduced into German intellectual discourse by Moses Hess and was used in diverse ways by various representatives of the Hegelian left to represent the future transcendence of the nineteenth-century European social order. But Marx's vision of communism differed from that of his contemporaries in ways that can be fully understood only once his grounding of time transcendence in time through proletarian revolutionary action is taken into account. As the concept of communism later played such a central role in debates leading up to Leninism and Stalinism, it is worth examining this point in some detail.

Most significantly, whereas many of the Young Hegelians had seen "communism" as a phenomenon that would exist sometime in the future, that would be realized "after" the current historical period was transcended, Marx emphatically argued that if communism was to be made practical, it had to be understood as something *present*. As he put it in the *1844 Manuscripts*, "*Communism* is the necessary form and dynamic principle of the immediate future, but communism as such is not the goal of human development—the form of human society."[16] The same point was reiterated, in even clearer form, by Marx and Engels in *The German Ideology*: "Communism is for us not a *state of affairs* which is to be established, an *ideal* to which reality will have to adjust itself. We call communism the *real* movement which abolishes the present state of things."[17]

To place communism on an abstract time line, to define it as a form of social order to be realized at some point in the short-term or long-term future, is thus to misunderstand its history-transcending nature. Communism cannot liberate humanity from time if its realization is confined to ordinary temporal categories. Instead, with the realization of communism, time-bound "history" itself is relativized. In both Marx's early and later writings, he tends to portray current empirical history as a kind of temporal illusion, as his references to it as "the *entire so-called history of the world*," "the history of all hitherto existing society," and "the prehistory of human society" suggest.[18] From Marx's point of view, communism—a revolutionary movement that exists in time and yet transcends ordinary time—is "the riddle of history solved, and it knows itself to be this solution."[19]

Marx's concept of communism thus represents his resolution of the paradoxes inherent in Hegel's conception of time. Like Hegel, Marx accepts the fundamental necessity of an empirical, rational understanding of human history as subject to temporal laws. Also like Hegel, Marx postulates the simultaneous necessity of some supratemporal force that provides history with meaning and direction, thus subordinating the concept of the rational time grid to a charismatic denial of time's inexorability. Unlike Hegel, however, Marx insists that the transcendence of history and time can be brought about through human action in the world. The mundane activity of labor demonstrates humanity's potential for overcoming ordinary time on a day-to-day basis. But this potential cannot be realized in capitalist society, where labor itself has become thoroughly alienated and impotent. Ultimately, only collective proletarian revolutionary action—communism—can reunify humanity's being in time with its potential for creative time transcendence. Communism is therefore the sole path toward human liberation.

Such a conception of communism may appear confusing or implausible to contemporary readers; indeed, it was quite difficult to fathom even for many of Marx's later disciples, who unlike Marx had not spent their formative years wholly immersed in problems of Hegelian philosophy. Yet an examination of Marx's later analysis of bourgeois society makes it clear that his philosophical writings of the 1840s remained central to his intellectual outlook throughout his life.

Indeed, Marx's grand economic critique of capitalism, to which he devoted the majority of his mature working years, is thoroughly imbued with the spirit of his early philosophical writings. Nowhere is this more evident than in Marx's exploration of the role of labor time in capitalist production. The "mature" Marx, no less than the "young" Marx, emphasized the creative role of human labor as a crucial element in his theoretical framework. As Marx writes in *Capital*: "Labor is, first of all, a process between man and nature, a process by which man, through his own actions, mediates, regulates and controls the metabolism between himself and nature. . . . In this way, he simultaneously changes his own nature. He develops the potentialities slumbering within nature, and subjects the play of its forces to his own sovereign power."[20] As in the *1844 Manuscripts*, Marx insists that creative labor allows human beings to realize a present intention in the future and thus to defeat the flux of time. Once again, Marx emphasizes that this is the decisive difference between people and animals:

We presuppose labor in a form in which it is an exclusively human characteristic. A spider conducts operations which resemble those of the weaver, and a bee would put many a human architect to shame by the construction of its honeycomb cells. But what distinguishes the worst architect from the best of bees is that the architect builds the cell in his mind before he constructs it in wax. At the end of every labor process, a result emerges which had already been conceived by the worker at the beginning, hence already existed ideally. Man not only effects a change of form in the materials of nature; he also realizes his own purpose in those materials.[21]

How, then, does the capitalist system lead to the exploitation, rather than the free enjoyment, of this potentially self-affirming activity?

The key to understanding the exploitative nature of capitalism, Marx argues, lies in its use of labor time. The value of any given commodity, according to Marx, can be traced ultimately to the "socially necessary labor-time" that goes into producing it;[22] this is the labor theory of value expounded by Adam Smith and David Ricardo before him. But Marx interprets the labor theory of value in a strikingly new way, one that logically flows from his conception of labor as, at least potentially, possessing a power greater than time. Since labor alone is the "living source of value,"[23] neither the mere exchange of commodities in trade, which presupposes that the things exchanged are necessarily of equal value, nor the increased use of capital in production in the form of advanced machinery can possibly generate profit. Profit, the very motor of the capitalist system, can exist only because labor itself, which is uniquely capable of generating new value, is also exchanged as a commodity on the open market. But the *price* of labor on the market reflects not its potential to create value but, rather, as with every other commodity in a capitalist economy, the labor time that is socially necessary for the "production" of laborers—that is, labor receives a subsistence wage. If a capitalist hires a worker for, say, ten hours of work, the worker may produce value equal to the cost of his subsistence in, say, five hours; this in effect means that the value produced by his labor in the subsequent five hours of the working day is a free gift to the capitalist. The sale of labor time as a commodity under capitalism, which disregards labor's special quality as the fundamental mode of human self-actualization vis-à-vis nature and its unique ability to create value, is thus for Marx the secret and sole source of capitalist "surplus value."

For this reason, the political struggle of the working class for a shorter working day becomes a crucial expression of communist consciousness. Just as the *Communist Manifesto* referred to the fight for the Ten Hours' Bill as a crucial example of the concrete achievements of class struggle, an entire chapter of *Capital* is devoted to the subject of the "struggle for a normal working day" in England.[24] Capitalists, Marx argues, will always tend to push the working day to its biological limits in order to maximize profit: "The capitalist likes nothing better than for [the worker] to *squander* his dosages of vital force as much as possible, without interruption."[25] A system of wages for commoditized labor measured in abstract units of "labor time" thus constitutes only a new form of slavery. Once the working class collectively realizes this, they will surely unite to overthrow the capitalist organization of "work time" and "free time." Clearly, control over time is a fundamental issue—perhaps *the* fundamental issue—in Marx's mature works of political economy.

Thus, while Marx's empirical focus and analytic mode of expression in *Capital* appear on the surface to mark a sharp break with the directly Hegelian concerns of his early writings, his central theme remains the same: to explain how free human control over daily time might emerge out of the heteronomous organization of socioeconomic time under capitalism. Hegel's philosophy had provided Marx with the charismatic interpretation of historical, linear time that lay at the foundation of his solution to this problem. The concept of communism as a collective movement for the realization of labor's potential for time transcendence within time transformed the Hegelian charismatic-rational conception of time into a revolutionary force. Finally, Marx's analysis of time use under capitalism demonstrated that the current organization of social life in abstract time contained inner contradictions that only became fully visible from the perspective of communist consciousness, which transcended existing understandings of time.

Political, Economic, and Cultural Dichotomies in Marx's Work

Marx's synthesis of charismatic and rational time orientations in the concept of proletarian revolutionary action is central to his whole theoretical project. The vision of communism provided Marx with a unique vantage point from which to analyze social development and to interpret the social arrangements of nineteenth-century Europe as only a transitory phase in a revolutionary historical process. There can be no denying the remarkable insights into modern capitalism generated from this intellectual starting point, many of which retain contemporary validity.

However, it is one thing to analyze an existing social order and another to propose strategies for building up a new one. In general, Marx avoided detailed descriptions of the "future society" that would emerge "after" the proletarian revolution; more specifically, as we have seen, he refused to see communism as subject to this sort of time-bound reasoning. Yet Marx did nonetheless take it upon himself fairly frequently to propose concrete strategies for furthering the communist movement. Believing that his own communist theorizing was only a reflection of the ongoing process of emancipated human labor coming into consciousness of itself, Marx did not hesitate to articulate what he felt were "the line of march, the conditions, and the ultimate general results of the proletarian movement."[26]

But in doing so, Marx unwittingly stumbled into a new set of theoretical contradictions, for only in the practical activity of revolution itself does Marx demonstrate how rational and charismatic time can be successfully synthesized. In his strategic prescriptions for realizing communism—politically, economically, and culturally—the two tend to fall apart again. Marx's theory of revolutionary politics simultaneously charts the gradual impact of the empirical proletarian struggle in time *and* prophesies a sudden break with all past forms of human society, making conventional time-bound politics irrelevant. Marx's analysis of the economics of socialism argues both for a more effective observance of norms of time discipline *and* for a complete transcendence of those norms. Finally, Marx's vision of communist culture argues that socialist human beings are both indelibly marked by history *and* radically freed from all historical constraints.

Let us begin with an examination of the dichotomy in Marx's conception of revolutionary politics. We have seen that while Marx conceives of labor as a potentially time-transcending force, he nonetheless denies that labor as it is presently constituted can free humanity from domination. Only in the process of proletarian revolution itself does the laboring class reclaim its potential for freely creative communal life. But how should political action be carried out when revolution itself is not imminent? In fact, Marx presents two wholly irreconcilable visions of political action—the first evolutionary, the second apocalyptic. Marx's evolutionary political strategy accepts linear historical time as binding on the proletarian revolutionary struggle, and counsels patience in order to make gradual progress within existing bourgeois institutions. Marx's apocalyptic political strategy calls for an immediate break with human "prehistory" through a revolutionary overthrow, not only of bourgeois society, but of rational time constraints on human action.

This contradiction is already apparent in the *Economic and Philosophical Manuscripts*, where it is expressed as a noticeable dualism in Marx's attitude toward the proletariat itself. The proletariat, as the class created exclusively by the modern economic organization of labor, is for Marx theoretically the key force for the overcoming of human oppression. But Marx is strangely ambiguous about the character of the proletariat as it currently exists, under the constraints of temporal economic laws. At first, Marx seems to insist that the effect of labor in modern industrial settings is to destroy all possibility of proletarian autonomy:

> It is true that labor produces wonderful things for the rich—but for the worker it produces privation. It produces palaces—but for the worker, hovels. It produces beauty—but for the worker, deformity. It replaces labor by machines, but it throws one section of the workers back to a barbarous type of labor, and it turns the other section into a machine. It produces intelligence—but for the worker stupidity, cretinism.[27]

But later in the same work, Marx describes the communal association of the existing proletariat as a harbinger of communist freedom—in Weberian terms, as a collective charismatic agent:

> When communist artisans associate with one another, theory, propaganda, etc., is their first end. But at the same time, as a result of this association, they acquire a new need—the need for society—and what appears as a means becomes an end. In this practical process the most splendid results are to be observed whenever French socialist workers are seen together. Such things as smoking, drinking, eating, etc. are no longer means of contact or means that bring them together. Association, society, and conversation, which again has association as its end, are enough for them; the brotherhood of man is no mere phrase with them, but a fact of life, and the nobility of man shines upon us from their work-hardened bodies.[28]

In the first passage, workers are depicted as automatons trapped increasingly and inexorably in productive slavery; in the second, they are portrayed as realizing their potential, here and now, to concretize the "brotherhood of man."

That Marx never quite decided between these two interpretations of his own argument is clear in his contradictory pronouncements about revolutionary strategy in capitalist Europe. In some writings, Marx places the gradual emergence of proletarian power in strictly time-bound, historical terms. Thus, in the 1859 preface to *A Contribution to the Critique of Political Economy*, Marx denies the independent role of "will" and "consciousness" in transforming the social order:

In the social production of their existence, men inevitably enter into definite relations, which are independent of their will, namely relations of production appropriate to a given stage of development of their material forces of production. The totality of these relations of production constitutes the economic structure of society, the real foundation, on which arises a legal and political superstructure and to which correspond definite forms of social consciousness. The mode of production of material life conditions the general process of social, political, and intellectual life. It is not the consciousness of men that determines their existence, but their social existence that determines their consciousness.[29]

From this perspective, attempts at time-transcending action are pointless, since "no social formation is ever destroyed before all the productive forces for which it is sufficient have developed."[30] Thus, Marx at one point argues that attempts to launch the proletarian revolution during periods of capitalist expansion are doomed to failure: "With this general prosperity, in which the productive forces of bourgeois society develop as luxuriantly as is at all possible within bourgeois relationships, there can be no talk of a real revolution. Such a revolution is only possible in the periods when both these factors, the modern productive forces and the bourgeois productive forms, come into collision with each other."[31] This interpretation of the process of revolutionary advance remains strictly bound to a notion of historical development in linear, abstract time.

Yet in other writings, Marx appears to be convinced that the proletariat is "mature" enough immediately to wipe out remaining vestiges of class society and transcend the constraints of historical laws. The *Communist Manifesto* itself concludes with a classic statement of this vision of revolutionary politics as a charismatic force for the collapsing of time: "The Communists turn their attention chiefly to Germany, because that country is on the eve of a bourgeois revolution that is bound to be carried out under more advanced conditions of European civilization, and with a much more developed proletariat, than that of England was in the seventeenth, and of France in the eighteenth century, and because the bourgeois revolution in Germany will be but the prelude to an immediately following proletarian revolution."[32] Here, Marx's emphasis is not so much on the slow development of objective material factors as on the degree of "civilization" and proletarian consciousness allowing for revolutionary progress; indeed, we are assured that in the storm of revolutionary struggle, the entire capitalist epoch can be collapsed into a short period.

Torn as it is between charismatic and rational time orientations, Marx's politics thus remains fundamentally contradictory. At times, Marx appears to think revolutionary action can accelerate the pace of history; at times, he argues that action at the wrong time can never succeed. Indeed, during Marx's lifetime, his advice to revolutionaries throughout Europe displayed remarkably little consistency on the question of how to "time" revolutionary action most effectively. Sometimes Marx counseled patience to give economic conditions time to mature; at other times he advocated immediate revolutionary action in places where capitalism had hardly begun to develop.

Such contradictions might appear unimportant to the prospects for socialism in light of Marx's claim that politics represents part of the "superstructure" and is therefore essentially determined by underlying economic forces. Yet the same ambiguity between rational and charismatic understandings of time is found in Marx's analysis of socialist economics. There are two distinct economic alternatives that might be derived from Marx's critique of capitalist exploitation: one based on the rational conception of time, and one based on a charismatic conception of socialism as beyond ordinary time constraints. Again, Marx inconsistently embraces both.

According to the first argument, capitalism must be abolished because, given the inherently limited nature of each person's time, it is manifestly exploitative for capitalists to extract the value-creating potential of a full day's labor—let alone extend that "full day" to eighteen hours in the most extreme cases—while only paying a wage equal to the value produced in a mere fraction of the working day.[33] Such an argument is essentially tied to a rational conception of time and a negative conception of labor. On this view, time is an abstract grid, the same for everyone; a system in which the leisure time of a few is made possible by the continuous drudgery of the majority clearly cannot stand.[34]

Although some element of mundane labor will remain necessary in socialist society, the main point in abolishing capitalism, from this point of view, is to expand the realm of free time relative to working time: "The realm of freedom really begins only where labor which is determined by necessity and external expediency ends; it lies by its very nature beyond the sphere of material production proper. . . . The true realm of freedom, the development of human powers as an end in itself, begins beyond it, though it can only flourish with this realm of necessity as its basis. The reduction of the working day is the basic prerequisite."[35] Labor here seems inextricably chained to the necessity of time; socialist society will therefore still have to organize labor time—the "realm of necessity"—in much the same way as under capitalist society. Indeed, time discipline under socialism, from this perspective, will, if anything, become even *more* important:

> On the basis of communal production, the determination of time remains, of course, essential. The less time the society requires to produce wheat, cattle, etc., the more time it wins for other production, material or mental. Just as in the case of an individual, the multiplicity of [society's] development, its enjoyment and its activity depends on economization of time. Economy of time, to this all economy ultimately reduces itself. Society likewise has to distribute its time in a purposeful way, in order to achieve a production adequate to its overall needs; just as the individual has to distribute his time correctly . . . in order to satisfy the various demands on his activity. Thus, economy of time, along with the planned distribution of labor time among the various branches of production, remains the first economic law on the basis of communal production. It becomes law, there, to an even higher degree.[36]

Here Marx presents a vision of socialism that is explicitly subject to the rule of time—and as a result hardly seems very inspiring as a vision of human liberation.

According to the second, charismatic argument, however, it is precisely the subjection of labor to bourgeois time constraints that makes capitalism exploitative—and therefore socialism must abolish abstract, rational time itself. Under capitalism, this argument runs, the free essence of labor, which potentially transcends time's dominion over man, is alienated from the worker and subordinated to the dead weight of capital. As Marx vividly puts it, "Capital is dead labor which, vampire-like, lives only by sucking living labor, and lives the more, the more it sucks."[37] On this view, the sole obstacle to human freedom is that the capitalist mode of production mandates the rule of "dead labor" over "living labor"; capitalism must be overthrown to reverse this relationship. As the *Communist Manifesto* puts it:

> In bourgeois society, living labor is but a means to increase accumulated labor. In Communist society, accumulated labor is but a means to widen, to enrich, to promote the existence of the laborer.
>
> In bourgeois society, therefore, the past dominates the present; in Communist society, the present dominates the past.[38]

From the charismatic perspective, then, economizing calculations of time use will be irrelevant after the victory of communism. Marx at one point even goes so far as to suggest that this increase in disposable time invalidates the labor theory of value itself under socialism:

> Capital . . . increases the surplus labor time of the mass by all the means of art and science, because its wealth consists directly in the appropriation of surplus labor time. . . . It is thus, despite itself, instrumental in creating the means of social disposable time, in order to reduce labor time for the whole society to a diminishing minimum, and thus to free everyone's time for their own development. But its tendency is always, on the one side, *to create disposable time, on the other, to convert it into surplus labor.* . . . The more this contradiction develops, the more does it become evident that the growth of the forces of production can no longer be bound up with the appropriation of alien labor, but that the mass of workers must themselves appropriate their own surplus labor. Once they have done so—and *disposable time* thereby ceases to have an *antithetical* existence—then, on one side, necessary labor time will be measured by the needs of the social individual, and, on the other, the development of the power of social production will grow so rapidly that, even though production is now calculated for the wealth of all, *disposable time* will grow for all. . . . The measure of wealth is then not any longer, in any way, labor time, but rather disposable time.[39]

It is not at all clear, from an economic point of view, how one might calculate "necessary" labor time in terms of the "needs of the social individual, or how basing the "measure of wealth" on disposable time would work in practice. But in passages such as these, Marx is no longer arguing from a rational time-bound perspective but from a charismatic, time-transcendent one; precise measurements of work time and disposable time will hardly be crucial once "disposable time . . . ceases to have an antithetical existence" to labor time. Indeed, since labor under communism becomes itself an expression of freedom, the distinction between "labor time" and "free time" itself quite naturally disappears: "It goes without saying . . . that direct labor time itself cannot remain in the abstract antithesis to free time in which it appears from the perspective of bourgeois economy. . . . Free time—which is both idle time and time for higher activity—has naturally transformed its possessor into a different subject."[40] With labor at last able to realize its inherent potential for transcending time, labor itself will become part of "free time"— which, under communism, is all the time.

We have seen that while Marx successfully synthesized charismatic and rational time orientations in the idea of proletarian revolutionary action, which collapses time within time, his views on political and economic organization in time remain fundamentally dichotomous, sometimes accepting and sometimes rejecting the need for time constraints under socialism. It remains to discuss the third dichotomy left unresolved within Marx's work—Marx's contradictory interpretation of postrevolutionary culture. Underlying Marx's vision of postrevolutionary social organization, and at the base of Marx's theoretical, political, and socioeconomic arguments more generally, there is a cultural ideal: a vision of human subjects who continually engage in fulfilling, time-transcending activity within the temporal world and thus express the creative "species-character" that distinguishes them from animals.

But how are time-bound proletarians, molded by class-based economic life, to be transformed into communist workers? As in the case of revolutionary politics and socialist economics, there are two possible answers here. From the rational temporal perspective, the cultural transformation involved in the proletarian revolution will necessarily be a gradual process in abstract, linear time, itself governed by "laws" of social development. From the charismatic perspective, by contrast, the revolution will have so radically changed the conditions of human social life that the laws that govern ordinary, prerevolutionary time will no longer be an obstacle to a universal free community.

Again, Marx's prescriptions for socialism remain fundamentally torn between these rational and charismatic interpretations of what should be done. This is strikingly illustrated in Marx's division of the postrevolutionary era itself into two periods, the "lower" and "higher" stages of communism—with the former based almost wholly on an abstract understanding of time, the latter on a charismatic ideal of time transcendence. The first stage of communist society, as Marx puts it, is "in every respect, economically, morally, and intellectually, still stamped with the birth-marks of the old society from whose womb it emerges"; thus, there is a linear historical relationship between this stage of communism and its capitalist predecessor.[41] Because of the continuing influence of the remnants of the bourgeoisie, and of bourgeois habits of economic "parasitism," the "dictatorship of the proletariat" governing this first stage of communism must make a number of concessions to time in order to survive. For example, Marx argues that during this period, workers' wages will be calculated on the basis of the formula "from each according to his ability, to each according to his work"—a formula that requires a calculation of the number of hours of "socially necessary" labor performed by each worker. Furthermore, certain deductions from the social wage fund as a whole must take place, such as a fund for replacement of the "wear and tear" on machinery over time.

The "higher" stage of communism, however, is based on the overcoming of all these time constraints. As a charismatic ideal, it is of course far less clearly defined in institutional terms than the "lower" stage of communism. However, Marx's basic outline of the "higher" stage of communism makes it clear that the concessions to time made in the "lower" stage are no longer necessary, because of a cultural transformation of human beings themselves. Labor itself has become "life's prime want" and the mainspring of abundant wealth, and external labor discipline is therefore an anachronism:

> In a higher phase of communist society, after the enslaving subordination of the individual to the division of labor, and therewith also the antithesis between mental and physical labor, has vanished; after labor has become not only a means of life but life's prime want; after the productive forces have also increased with the all-round development of the individual, and all the springs of co-operative wealth flow more abundantly—only then can the narrow horizon of bourgeois right be crossed in its entirety and society inscribe on its banner: From each according to his ability, to each according to his needs![42]

From this perspective, communism is indeed "the riddle of history solved";[43] communist culture is not to be thought of as subject to ordinary historical constraints. Indeed, as Booth has emphasized, there appears to be a distinct lack of temporal trade-offs in Marx's vision of daily life under communism. As Marx once famously put it: "In communist society, where nobody has one exclusive sphere of activity but each can be accomplished in any task he wishes, society regulates the general production and thus makes it possible for me to do one thing today and another tomorrow, to hunt in the morning, fish in the afternoon, rear cattle in the evening, criticize after dinner, just as I have a mind, without ever becoming hunter, fisherman, shepherd or critic."[44] Despite Marx's desire to remain faithful to the scientific method—and to the abstract, rational conception of time upon which, in his day, modern science was thought to depend—in descriptions of communist culture such as these he articulates a purely charismatic vision of a society beyond temporal constraints, and thus beyond economic scarcity.

Marx's vision of communism appears ultimately to be an inadequate response to the Kantian and Hegelian dilemma of how to reunite human empirical existence with time-transcendent meaning, for in Marx's discussions of revolutionary politics, of socialist economic organization, and of communist culture, the dichotomy between charismatic and rational time orientations reappears. Revolutionary politics depends simultaneously upon the heroic action of the proletariat to conquer time and the subordination of politics to time-bound processes of economic development. The socialist economy must both transcend the "bourgeois antithesis between labor and leisure time" *and* "economize time to an even higher degree" than capitalism. On a cultural level, proletarian revolution is supposed to wipe out the "muck of ages" and purify the proletariat to "form society anew"—yet communist revolutionaries are nonetheless compelled to organize institutions to deal with the historical legacy of capitalism for a lengthy cultural transition period.

Since a final revolutionary victory over time appears impossible, there is an implicit tendency in Marx's work to argue instead for a disciplined, revolutionary movement against time *without end*. Revolutionary action here and now—participation in the movement that abolishes the present—is the most direct possible empirical expression of one's belief in the human possibility of transcending historical constraints and of building a world beyond current understandings of time. This was the central message of Marx's theoretical response to Hegel's tragic view of history, which had reconciled a charismatic transcendence of time with empirical, time-bound history only at the cost of negating the ultimate importance of human action in the temporal world. Revolution, for Marx, reasserts human autonomy in a "practical movement"[45] that empirically reconciles human beings' temporal existence with their timeless essence, their subordination to the realm of necessity now with their participation in a world of freedom under communism. As the later history of Marxism shows, "faithful" Marxism thus required more than an understanding of the historical role of the proletariat. It required direct participation in proletarian revolutionary action as well—even by the theoretically conscious bourgeois intelligentsia, which, according to Marx and Engels, at "the decisive hour . . . cuts itself adrift and joins the revolutionary class, the class that holds the future in its hands."[46]

Engels and the Institutionalization of Theoretical Marxism

Unlike the death of Hegel, Marx's death in 1883 did not create an immediate "succession problem," since one of the authoritative founders of "scientific socialism," Friedrich Engels, survived until 1895. Because of his intimate, lifelong collaboration with Marx, Engels's legitimacy as his spokesperson was never seriously questioned within the emerging Marxist movement. Engels himself appeared to relish this role, doing more to popularize Marx's interpretation of socialism in the twelve years that remained to him than Marx himself had done in a lifetime. However, close analysis reveals that Engels's popularization of Marxism was not entirely in keeping with the spirit of Marx's own work. Since Engels's later writings were influential among later Marxists such as Bernstein, Kautsky, and Lenin, it is worthwhile to examine them briefly in light of the argument so far.

Interpretations of the Marx-Engels relationship in the literature range from the ritual assertion by Soviet scholars—tacitly accepted by many Western analysts—that Marx and Engels were of one mind on every significant issue to the contrary claim that Engels's representation of Marxism was a "tragic deception" that transformed Marx's nuanced theory into a dogma and thus set the stage for Stalinist totalitarianism.[47] If the former interpretation leads to an unwarranted underestimation of Engels's independent role in the formation of the Marxist movement, the latter view wildly overestimates the influence of Engels's work on later generations of Marxists, who were, after all, perfectly capable of reading and interpreting Marx's own writings for themselves, even if Engels's summaries were what first motivated them to do so. The most careful analyses of this issue adopt an intermediate position: although Engels's partnership with Marx in the 1844–48 period appears to have been built on a genuinely fruitful interchange between two like-minded former left Hegelians, after 1859 Engels became increasingly convinced that Marx's "dialectical method" could be systematized as a universal explanation of all scientific phenomena, social and natural—a project that Marx did not directly discourage but for which he person-ally had little enthusiasm.[48]

Engels's attempts to show how various scientific discoveries in Dar-winian evolutionary biology or in the study of electricity verify the truth of the dialectical method—summed up by Engels in the axiom that all existence is nothing more than the "motion of matter"—are hardly worthy of serious intellectual scrutiny.[49] More important from the per-spective of this study, the idea that Marx's dialectic of labor was some-how a subset of a larger "dialectic of nature" represented a subtle trans-formation of Marx's conception of time, for Marx's central notion of human labor as a potentially time-transcendent force that could be actualized in mundane time through revolutionary struggle could not be logically maintained if one simultaneously insisted that labor was essentially biological and therefore wholly subject to deterministic natural laws. Whereas Marx's vision of communism synthesized the charisma of social action with the rational analysis of empirical social conditions, Engels increasingly portrayed socialism as simply the fulfillment *within linear time* of a predetermined historical destiny, unfolding according to the "dialectical" process that Marx had been the first to discover.

Indeed, in his first and most influential effort to present Marxism as a "system" of philosophical and scientific knowledge, the *Anti-Dühring*, Engels insisted that the belief that anything could transcend time was absurd: "The basic forms of all being are space and time, and being out of time is just as gross an absurdity as being out of space."[50] Ridiculing Dühring's contention that time was nothing more than a name given to change within concrete existence, Engels insisted that "just because time is different from change, is independent of it, it is possible to measure it by change, for measurement always requires something different from the thing to be measured. And time in which no recognizable changes occur is very far from *not* being time; it is rather *pure* time, unaffected by any foreign admixtures, that is, real time, time *as such*."[51] But if "real time" is thus in a sense ontologically prior to all the changes that take place within it, then how can human action possibly be an expression of freedom—of the ability to make one's time one's own?

Ultimately, Engels's attempt to embrace a quasi-Newtonian conception of "scientific," rational time while still insisting on the inevitable progress of human beings toward freedom under socialism resulted in a view of world history that was more Hegelian than Marxian—with the "dialectic" now playing the role of *Geist*. Engels even drew upon Hegel, rather than Marx, for his definition of freedom itself:

> Hegel was the first to state correctly the relation between freedom and necessity. To him, freedom is the insight into necessity. . . . Freedom does not consist in any dreamt-of independence from natural laws, but in the knowledge of these laws and in the possibility this gives of systematically making them work towards definite ends. . . . Therefore the *freer* a man's judgment is in relation to a definite question, the greater is the *necessity* with which the content of this judgment will be determined.[52]

The direct contradiction between Engels's formula that "freedom is the insight into necessity" and Marx's definition of freedom as possible only where "labor which is determined by necessity and mundane considerations ceases" is striking. One is tempted to interpret Engels's argument here as an unconscious product of his strict Calvinistic upbringing; indeed, his later theoretical works present an implicit doctrine of salvation through the embrace of socialist "predestination" that increasingly overshadowed the call for organized revolutionary struggle.

Engels's codification of Marxism thus represents an institutionalization of Marx's charismatic-rational vision of communism as a kind of theoretical orthodoxy, requiring followers to replicate the sacred analytic techniques of the founders in order to obtain historical redemption. In Engels's presentation of Marxism, time reasserted itself as an inexorable force; yet efficacious action within time nonetheless remained possible through the charisma of Marx's "method" of "dialectical materialism." Faith in "Marxism" thus replaced direct participation in revolutionary action as the main test of membership in the socialist community. Ironically, this transformation made it much easier to recruit and train a new generation of "scientific socialists" in a period when the experience of direct struggle with actual workers was becoming increasingly rare among left-wing intellectuals. It was on the foundation of an insistence on the unshakable "truth" of Marxism that the Second International—founded in 1889 by Engels and the leaders of the German Social Democratic Party (SPD)—was built.

However, in claiming that future socialist salvation could be obtained only through fidelity to Marx's "system," Engels did not wholly obliterate the revolutionary elements of Marx's worldview. The question of how one should define "fidelity" to Marx in political practice remained ambiguous in Engels's work, as it had in Marx's original writings. Engels's systematization of Marxism helped strengthen the faith of younger Marxists in dialectical historical progress toward socialism; but how to participate in that progress empirically was still a matter of debate.

Marxism in the Second International

Marx's work, I have argued, transformed the charismatic-rational time orientation first developed in Hegel's philosophy into a theoretical justification for collective action to transcend time within the confines of empirical time. As we shall see, this theory later became the basis of the patterns of time use institutionalized by Leninist polities and Stalinist economies. Indeed, developments in Marxism shortly after the deaths of Marx and Engels parallel later developments after the deaths of Lenin, the founder of the Soviet political system, and Stalin, the founder of the Soviet socioeconomic system. Therefore, the period between the creation of Marx's social science and the Russian Revolution of 1917 can itself be seen as a cycle within the institutional development of the charismatic-rational conception of time—the initial theoretical cycle.

Marx left his theoretical disciples a complex legacy. Having synthesized the linear, abstract conception of time at the core of Newtonian science with the possibility of charismatic mastery over time through proletarian revolutionary action, he nonetheless failed to provide any clear guidelines on how to maintain fidelity to both sides of the charismatic-rational dualism in the building of effective socialist political institutions, in the creation of a workable socialist economy, or in the building of a communist culture. Of course, had Marx's predictions of an imminent capitalist collapse and the subsequent realization of communist society been correct, none of this would have mattered.

From the Weberian perspective adopted here, however, the social impact of any purely theoretical revolutionary vision depends upon whether it can provide some basis for the welding together of the ideal and material interests of a significant group of people into institutions they perceive to be legitimate. If a revolutionary theory lacks this ability, the revolutionary movement based on that theory will lack the power to withstand the inevitable routinization of initial revolutionary enthusiasm. Thus, for a new regime type to emerge out of a revolutionary movement, a successful political stage of institutional development must follow the theoretical stage.

The question of how best to translate Marxist theory, as systematized by Engels, into a practical political force was, in fact, at the center of debate within the Second International. By the 1890s, with the growth of the German SPD into a major political force commanding the allegiance of millions of workers, and with the increasingly explicit self-identification of the party leadership with Marxism as the syndicalist Ferdinand Lassalle's original influence receded, the difficult and divisive task of constructing a detailed, positive Marxist political program was placed unavoidably on the agenda. On this issue, three basic alternatives were predominant within the Second International: the revisionism of Eduard Bernstein, the revolutionism of Rosa Luxemburg, and the orthodoxy of Karl Kautsky. As in the period after Hegel's death, then, the charismatic-rational synthesis presented in the work of "founders" Marx and Engels routinized into "right," "left," and "center" factions.

As in the Hegelian case, these divisions reflected three possible interpretations of the degree to which the reconciliation of charismatic time
transcendence and rational time measurement could be said to have
been realized historically. But whereas for disciples of Hegel this question was put in terms of the degree to which *Geist* had manifested itself
in current social arrangements, for fin de siècle Marxists the analogous
problem became one of assessing the degree of revolutionary potential
of the present-day proletariat. "Leftists" believed that the proletariat
was fully capable of fulfilling its historical destiny immediately and that
any attempt to delay the communist revolution was therefore inherently
un-Marxist. "Right" Marxists insisted that the present-day working
class was empirically uninterested in revolution and that a long evolutionary process would be necessary to realize socialism; they saw calls for
immediate revolution as untrue to the rational, scientific components of
Marxism. Finally, "center" Marxists considered the raising of proletarian "communist consciousness" to be an ongoing dialectical process, the
success of which depended upon remaining simultaneously faithful to
both the revolutionary and the scientific sides of Marx's work.

The direct political consequences of holding each of these three positions within Marxism were also rather different from those associated
with left, right, and center Hegelianism. Whereas left Hegelians had
generally identified the embrace of radical democracy and atheism as
the key to overcoming human alienation, left Marxists claimed that
only in revolutionary struggle itself did human beings realize their
inherent powers of collective self-creation. Whereas right Hegelians
argued (as had Hegel himself) that the self-realization of *Geist* in history validated the Prussian monarchy and Protestantism, right Marxists
claimed that the development of a modern working class would guarantee slow, continual progress toward socialism within the context of linear, abstract time. Finally, whereas center Hegelians had tried to medi-
ate between reactionary monarchism and militant opposition to the
status quo by supporting gradual liberal reform within the Prussian
context, center Marxists tried to synthesize proletarian revolution with
analytic social science by calling for "fidelity" to Marx's doctrine as the
guiding principle of the Social Democratic Party itself.

In Weberian terms, the charismatic-rational conception of time, as
institutionalized on the theoretical level by Marx and Engels, under-
went a complex but remarkably logical form of routinization. Right
Marxists legitimated their authority within the movement in rational-
legal terms, claiming to be scientific analysts who preserved democratic

procedures in their efforts to help the working class. Left Marxists legitimated their authority in charismatic terms, as personal representatives of the suprahistorical, sacred force of proletarian revolution—a form of authority that was itself subject to further routinization in nonrevolutionary periods. Finally, center Marxists attempted to legitimate their authority within the movement on what might be termed "neotraditional" grounds—by establishing a direct "lineage" from the great founders, Marx and Engels, to the current leadership of an essentially bureaucratic Social Democratic Party.[53]

These right, left, and center tendencies of Marxism within the Second International were far from clearly defined at first. In fact, as long as Marxists generally remained excluded from the existing political institutions of bourgeois Europe, different interpretations of Marxism appeared to have far more in common than was actually the case. Both those who placed more emphasis on the scientific unfolding of capitalism's self-destruction in time and those who emphasized the need for revolutionary action to destroy time could arrive at the same orthodox conclusion: the days of the capitalist system were numbered, and it was enough simply to identify with the historically progressive proletariat in the developing struggle to prove one's revolutionary credentials.

However, with the publication by Eduard Bernstein in 1896–99 of a series of articles on "evolutionary" Marxism in the Social Democratic Party's key theoretical journal, *Die Neue Zeit*, the question of how to reconcile the political institutionalization of Marxism with its revolutionary identity became a burning issue within the movement.[54] Bernstein, who had been in previous years a close associate of Engels and a passionate defender of orthodox Marxism, was perhaps the last person that could have been expected to provide a codification of the tenets of a new reformist interpretation of the doctrine. Bernstein himself denied any conflict between his views and what he took to be the essence of Marx's theory. However, in his revision of Marx's doctrine, Bernstein discarded practically every charismatic element in favor of a rational, evolutionary theory of social democracy.[55]

The 1899 introduction to Bernstein's book, in fact, brought the issue of time immediately to the forefront:

The theory which the *Communist Manifesto* sets forth of the evolution of modern society was correct as far as it characterized the general tendencies of that evolution. But it was mistaken in several specific deductions, above all in the estimate of the *time* the evolution would take. . . . But it is evident that if social evolution takes a much greater period of time than was assumed, it must also take upon itself *forms* and lead to forms that were not foreseen and could not be foreseen then.[56]   •

A truly scientific examination of Marx's theory was therefore necessary if residual "Utopian" elements of Marxism were to be excised from the doctrine. Here Bernstein, like Engels, embraced a thoroughly Newtonian conception of science as the progression of knowledge within linear time: "With advancing knowledge, propositions to which formerly absolute validity was attached are recognized as conditional and are supplemented by new scientific propositions which limit that validity, but which, at the same time, extend the realm of pure science."[57] Unlike Engels, however, Bernstein applied this sort of rational scrutiny to the principles of "dialectical materialism" themselves. From the rational perspective, Bernstein argued, it would have to be recognized that Marx's propositions about the end of class antagonisms and about the ultimate victory of socialism were "not capable of proof" but only "hypotheses more or less founded." Moreover, such prophetic elements were "not essential to the theory," at least in scientific terms.[58]

In Bernstein's view, the proletariat, rather than being identified as an inherently revolutionary class, had to be seen simply as an empirical group of individuals whose real interests lay in increasing their material standard of living within the capitalist framework: "We have to take working men as they are. And they are neither so universally pauperized as was set out in the *Communist Manifesto*, nor so free from prejudices and weaknesses as their courtiers wish to make us believe."[59] Accordingly, Bernstein recommended that the SPD, in order to broaden its electoral appeal, drop its exclusive concern with the working class and begin to take into account the interests of the peasantry and petty bourgeoisie as well. Moreover, Bernstein argued that capitalism had overcome its early history of periodic crises through innovations such as the credit system; accordingly, his work rejected any notion that socialism would represent a decisive break in history and argued instead that socialism would evolve out of the existing capitalist framework. Finally, Bernstein's embrace of rational time eliminated any possibility of a radical change in the relationship of human beings to labor time under some future communistic organization of economic life: "Of a general reduction of hours of labor to five, four, or even three or two hours, such as was formerly accepted, there can be no hope at any time within sight, unless the general standard of life is much reduced. Even under a collective organization of work, labor must begin very young and only cease at a rather advanced age, [if] it is to be reduced considerably below an eight-hours' day."[60] The rejection of Marx's notion of labor as a potentially time-transcending force here could hardly be more explicit.

In adumbrating this doctrine, Bernstein did draw on hints from within Marx's writings. Marx himself insisted on the necessity of struggles for reform within the capitalist system and even suggested that in certain circumstances the transition to socialism could be effected peacefully at the ballot box.[61] However, Bernstein divested Marx's theory of the crucial revolutionary component that gave Marx's arguments their particular cast. For Marx, the struggle for reforms under capitalism was only a training ground for the ultimate conquest of political power by the working class, and the possibility of a peaceful transition to socialism was combined with an insistence on the need for a subsequent "revolutionary dictatorship of the proletariat" over surviving representatives of the bourgeoisie.[62] Bernstein, by contrast, argued that socialism and capitalism were connected through evolution rather than separated through revolution and that "catastrophic" social change would in fact be a disaster for the working class.[63]

This tendency was succinctly expressed in what became for orthodox Marxists the most notorious slogan of Bernstein's revisionism: "That which is generally called the ultimate aim of socialism is nothing, but the movement is everything."[64] This slogan bears a superficial resemblance to Marx's statement that "communism is . . . not a *state of affairs* which is to be established, an *ideal* to which reality [will] have to adjust itself," but "the *real* movement which abolishes the present state of things." However, whereas Marx argues for the revolutionary *abolishing* of the present, Bernstein insists on the *acceptance* of the present and opposes any "final aim" that lies beyond rational time constraints: "I have at no time had an excessive interest in the future, beyond general principles; I had not been able to read to the end any picture of the future. My thoughts and efforts are connected with the duties of the present and the nearest future, and I only busy myself with the perspectives beyond so far as they give me a line of conduct for suitable action now."[65]

As the implications of Bernstein's revisionism for the self-conscious identity of the Marxist movement as *both* politically effective *and* revolutionary became clear, a nearly universal chorus of condemnation rained down upon Bernstein's head. This is all the more striking in view of the fact that as far as the first of these criteria—political effectiveness—is concerned, Bernstein's revisionism had much to offer. Indeed, a generally reformist rather than revolutionary tendency within the SPD had already become manifest by this time as a result of the party's need to appeal to an increasingly complacent electorate within a prosperous imperial Germany. Only the intense identification of the party leadership with the other, revolutionary strand within Marxism can explain the vehemence of the attacks on Bernstein for making explicit what had become the implicit policy of the party.

A second tendency within the SPD and the Second International became apparent a few years after the denunciation of Bernstein and revisionism—"left" Marxism. During the attack on Bernstein, of course, what later became the left and the center were fused in their common opposition to the denigration of the charismatic aspects of Marxism implicit in Bernstein's work. But as it became clear that the majority of the SPD leaders were content to denounce reformism in words without combating it in practice, a group emerged that was as committed to fighting the complacency within the SPD orthodox leadership as it was committed to the struggle against Bernstein's explicit views. This group found its most effective and representative spokesperson in Rosa Luxemburg, whose writings taken as a whole can be seen as the left Marxist equivalent of Bernstein's *The Premises of Socialism and the Tasks of Social Democracy*.

Luxemburg's first major work, in fact, was a forceful attack on Bernstein's views entitled *Reform or Revolution?*[66] But while Luxemburg herself recognizes in this essay that from a strictly Marxist viewpoint these two terms should not be considered as being in opposition, she soon slides into the tendency to dichotomize them—favoring "revolution" *over* "reform"—implied by her title. Indeed, Luxemburg eventually declared her adherence to a formula more or less the contrary of Bernstein's: "The movement as an end in itself, unrelated to the ultimate goal, is nothing to me; the ultimate goal is everything."[67] Despite its more revolutionary emphasis, this formula, like Bernstein's, postulates a sharp separation between the evolutionary, rational component of Marxism and its revolutionary, charismatic component, rather than arriving at any synthesis of the two.

This separation, in the end, could not help but lead Luxemburg to a conclusion precisely the opposite of Bernstein's: whereas Bernstein argued for political effectiveness at the expense of revolutionary identity, Luxemburg opted for a defense of revolutionary identity with a much less workable strategy for political effectiveness. For Luxemburg, the only cure for bureaucratism and complacency in the socialist movement, and the spur to revolutionary success, was mass spontaneous action by the proletariat. Her faith in the revolutionary character of the masses, and her hatred of those who would try to dictate the "proper" methods of revolutionary behavior to them, are clear in her essays discussing the mass strike as a revolutionary strategy. During a mass strike, she argues,

> there are quite definite limits set to initiative and conscious direction. During the revolution it is extremely difficult for any directing organ of the proletarian movement to foresee and to calculate which occasions and factors can lead to explosions and which cannot.... In short, in the mass strikes of Russia the element of spontaneity plays . . . a predominant part, not because the Russian proletariat are "uneducated" but because revolutions do not allow anyone to play the schoolmaster with them.[68]

Along with her tendency to use the terms "mass strike" and "revolution" interchangeably in this work, which makes the revolution itself appear to be a spontaneous event, Luxemburg assumes that once the revolution begins, each and every proletarian will naturally act in a way that denies the importance of temporal concerns: "In the storm of the revolutionary period even the proletarian is transformed . . . into a 'revolutionary romanticist,' for whom even the highest good, life itself, to say nothing of material well-being, possesses but little in comparison with the ideals of the struggle."[69]

The question immediately arises, however, What if this assessment of the spontaneous revolutionary potential of the proletariat is incorrect? If a "revolutionary storm" is approaching in which the masses of the oppressed will engage in acts of revolutionary transcendence, then indeed there might be no contradiction between revolutionary identity and political effectiveness. But what happens if the Bernsteinian hypothesis is correct and proletarians show no inclination to elevate "the ideals of the struggle" over the safeguarding of "life itself"?

The answer of the orthodox center—wait a little longer, then conditions may become ripe—was unacceptable to Luxemburg, who had a keen sense for the crucial revolutionary component of Marx's thought. As her battle with the complacency of the SPD leadership became more intense, and as she became increasingly isolated from the German Social Democratic mainstream, her conviction that revolutionary action was the only solution to the dilemmas of social democracy became even more pronounced. The disgrace of SPD support for war credits for imperial Germany in 1914 only sharpened these convictions, as did the outcome of the Russian Revolution in 1917—although Luxemburg criticized the Bolsheviks for their reliance on terror and their downplaying of the purifying and creative role of the revolutionary masses.[70] This drama, in which Luxemburg's passion for revolutionary transcendence began more and more to outweigh any considerations of tactical effectiveness, was played out to its seemingly inexorable conclusion: Luxemburg was executed, along with Karl Liebknecht, for her role in the abortive Spartacist revolution in Germany in 1919.

Thus, neither the reliance on evolution in time implied by Bernstein's reformism nor the denial of time constraints implicit in Luxemburg's strict revolutionism proved successful in creating an effective political movement that could weld together the contradictory elements of rationality and charisma within Marxism. Having dichotomized the alternatives of reform or revolution, each was pushed by a hidden dynamic toward a more complete reliance on one alternative and a more complete rejection of the other. Bernstein, in the long run, became a practical social reformer with little interest in or use for revolutionary breakthroughs, and Luxemburg became a revolutionist whose practical social analysis became less and less of a constraint on her desire for action. Each more nearly approached the ideal-typical conception of time that they leaned toward initially: Bernstein placed his hopes for reform on a long-term temporal process of social maturation that fundamentally accepted the rational view of time as inexorably linear and abstract; Luxemburg began to argue single-mindedly for an immediate charismatic revolutionary transcendence of existing historical conditions that ultimately had more in common with anarchism than with Marxism.

A third theoretical alternative within the SPD and within the Second International as a whole, therefore, won out—simple maintenance of doctrinal purity within the party leadership and the advanced sectors of the working class, combined with a faith that the dialectic of history would take care of the problem of socialism's further advance. This "center" tendency was represented above all by Karl Kautsky, a theorist skilled at pointing out the deviations of both right and left from orthodoxy but with little sensitivity to the need for an explicit strategy to bring about revolutionary change.[71] Kautsky, who had also begun his career as a socialist as a disciple of Engels, was, like his mentor, extremely interested in the possibility of giving Marxism a basis in natural science—especially Darwinism. Also like Engels, Kautsky understood Marxism to be an integrated system of knowledge that could potentially be applied to all realms of human understanding and that prophetically illuminated the future fate of the human species.

However, Kautsky was also willing to recognize explicitly something that was never faced directly by Marx or Engels: that a complete understanding of Marxist "science" required a degree of education and training well beyond the abilities of the vast majority of proletarians. The proposal that a theoretically advanced and orthodox Social Democratic Party leadership work to combat the "spontaneous" impulses of an insufficiently cultured working class was thus Kautsky's contribution to the political institutionalization of Marxism—not Lenin's, as countless commentators have assumed.[72] As Kautsky wrote in 1902, "The bearer of science is not the proletariat but the bourgeois intelligentsia. . . . Thus, socialist consciousness is something brought in from outside into the class struggle of the proletariat, and not something arising spontaneously out of it."[73] The main purpose of a Marxist politics would thus be to help the working class "learn" the truth of socialism, utilizing its experiences of exploitation within capitalist society as teaching tools. From this point of view, deviations from orthodoxy within the party, whether of the left or the right variety, were doubly damaging; not only did they undermine the cohesion of the socialist intelligentsia, but they sowed confusion among the less conscious elements of the proletariat that the party was supposed to direct. Despite his personal friendships with both Bernstein and Luxemburg, the "pope of Marxism" ultimately did his best to isolate them from the Social Democratic mainstream.

But Kautsky nonetheless retreated from the obvious practical question raised by his success in establishing the theoretical basis for a successful, mass-based socialist party: when and how could this party launch an actual socialist revolution? As he succinctly put it, the SPD should be seen as "a revolutionary party, not a revolution-making party."[74] Until the final crisis of capitalism came, the policy of an orthodox Marxist party could only be, as one analyst has described it, one of "revolutionary waiting."[75] And as the waiting continued, the revolutionary essence of the party was inevitably lost. The SPD became simply a German labor party—albeit a very successful one—and the other European parties of the Second International followed a similar path. The breakup of the Second International into national units with the outbreak of the First World War merely concretized what was already a de facto reality—revolutionary Marxism as an international movement had ceased to exist.

Thus, in addition to the inherent dualism in Marx's view of time, a further paradox became clear: Marx's combination of rational analysis of temporal processes and revolutionary time transcendence, which lay at the core of his attempt to reunite theory and practice, would *itself* become a mere theory increasingly remote from practice if no way were found to advance Marxism's political claims directly. A new phenomenon in the history of social theory emerged: one could not simply agree with everything Marx said and remain faithful to Marx. A doctrine committed to changing the world, and for which true change meant the emergence of a social order beyond conventional understandings of time, had both to change with the times in practice and to assert the time-transcending nature of the revolutionary cause it represented. Kautsky's "revolutionary waiting" only proved Bernstein's point: orthodox Marxism, despite its pretensions, could not overcome time. Nor did Luxemburg's activism provide a successful answer to the dilemma. What was needed for Marxism to remain an effective revolutionary force was a *welding together* of the charisma and rationality within Marx's theory—not a rejection of one or the other—within an effective political institution. Such an institution would not be created within Germany, however, but in Russia, under the leadership of Vladimir Il'ich Lenin.

4

THE POLITICAL CYCLE
From Lenin to the End of the NEP

This chapter examines Lenin's conception of time, its relationship to Marxism, and its influence on the formal institutional design of the political structures of the Bolshevik Party and Soviet state. Such an analysis must necessarily deal with several potential objections at the outset. To begin with, the notion that Lenin can be understood as a Marxist theorist contradicts the remarkably persistent tendency in the literature to portray his interest in socialism as merely instrumental, masking a more fundamental, unprincipled drive for power.[1] In Louis Fischer's words, "Lenin hewed as closely as necessary to the Marxist line: when life demanded change he deviated."[2] John Plamenatz expands on this theme: "Lenin left Marxism poorer than he found it; he took little interest in what was truly profound or subtle in it, and used it mostly as a source of quotations and slogans to justify his many and not always consistent policies. He did not even understand how he had distorted it."[3] Leonard Schapiro's assessment is much the same: "Lenin and his followers were less interested in doctrine than in preserving their own rule."[4] Even the recent biography by Dmitri Volkogonov— himself a self-professed former true believer in Marxism-Leninism—cites as Lenin's "main characteristic" his "radical pragmatism."[5]

Despite the rather widespread acceptance of this interpretation of Lenin's life and work, there are several good reasons to call it into question. To begin with, the idea that Lenin was simply a consummate pragmatist makes much more sense in hindsight than it would have in the pre-1917 period. Certainly no one would have advised an ambitious teenager in late-nineteenth-century Russia that the surest path to power was to join the revolutionary underground and write detailed treatises arguing for a strict application of Marx's materialist conception of history to Russian conditions. Within the narrower context of Russian Social Democracy, it is true, Lenin's ideological clarity, organizational ability, and insightful social analysis initially propelled him into a position of prominence. But even here Lenin's continuing insistence on the correctness of his distinctive interpretation of Marxism soon alienated all but a small core of loyal followers. Surely a true "pragmatist" would have found some way to compromise to obtain wider support!

In any case, it should be noted that many scholars who insist on Lenin's instrumental use of Marxism do not bother to examine Marx's works or those of his disciples in any detail. Such an examination makes it quite clear that the theoretical problems with which Lenin continually wrestled—how quickly society might move from bourgeois to proletarian revolution, where global proletarian revolution should begin, how the revolutionary movement itself should be organized—were precisely the political questions left unresolved by Marx and Engels and thereafter debated by all the Marxists of the Second International.

A second objection to taking Lenin's Marxism seriously, which in some cases overlaps with the first, is the argument that Lenin's "adaptation" of Marx's theory to Russian conditions is essentially much more a product of his Russianness than of his concern with socialism. Thus Theodore Von Laue refers to Lenin as "a secret Slavophile within the Marxist ranks."[6] For Liah Greenfeld, Lenin's advocacy of proletarian internationalism only serves to "mask" the "national sentiment behind it."[7] This theme has been developed most fully by Richard Pipes, who sees both tsarist autocracy and communist totalitarianism as expressions of a fundamentally authoritarian Russian cultural tradition. Indeed, Pipes goes so far as to conclude that "the less one knows about the actual course of the Russian Revolution, the more inclined one is to attribute a dominant influence to Marxist ideas."[8]

However, this interpretation, too, is difficult to square with all of the historical evidence. Of course Lenin grew up in Russia, read the works of Russian revolutionaries, and absorbed important cultural values from the Russian social environment. Yet if this is enough to qualify Lenin as a Russian nationalist, it would be difficult to find any political leader in the modern era that could not be so labeled. What is comparatively more striking is Lenin's remarkably consistent emphasis, both before and after the Bolshevik seizure of power, on the need to place Russian socialism in its proper international context. Lenin may have applauded Chernyshevsky's uncompromising revolutionism, but in his theoretical writings his consistent points of reference were the Germans Marx, Engels, and Kautsky. Lenin may have longed for his homeland during his days in exile, but the fact remains that he spent almost twenty of his fifty-three years immersed in the theoretical and social environment of urbanized, industrialized western Europe. He may have believed fervently in the possibility and even necessity of beginning the world revolution in Russia, yet after 1917 he devoted much of his attention— and a significant amount of the party's limited resources—to promoting the prospects for socialist revolution abroad.

It would be foolish to deny the existence of institutional and cultural continuities between the tsarist and Soviet regimes. But in some basic sense all social history is continuous. What marks the Leninist regime established in November 1917 as a qualitatively new phenomenon in Russian and world history—an identifiable regime type—is the fact that the Bolsheviks immediately began to build novel institutional structures, with a novel basis of legitimation, in the territories they controlled. Interpretations of Lenin as merely a disguised Russian nationalist make the nature of these institutional innovations quite difficult to explain.

A final, rather different objection to the study of the influence of Lenin's Marxism on the design of Soviet institutions is one generally associated with revisionist and structuralist interpretations of the Russian Revolution. Specifically, revisionists argue that Lenin and the Bolshevik elite simply did not have that much influence over the revolution's course and ultimate outcome. According to this interpretation, the true source of the breakdown of tsarism, of the collapse of the Provisional Government, and of the creation of a new, powerful Soviet state lies in structural social forces—particularly the contradiction between the feudal structure of agricultural production and the tsarist state's need to raise revenues for military expansion—that Lenin was powerless to direct, much less create.[9] In the cities, others have argued, the decisive role in the revolutionary year 1917 was played by the radicalized working class and soldiers, not the intelligentsia.[10] In Tim McDaniel's view, the uncompromising radicalism of the Bolshevik intelligentsia and the Russian proletariat alike can themselves be explained by the peculiar autocratic form of industrialization pursued by the tsarist state under Nicholas II; Lenin's version of Marxist ideology is from this perspective not an independent causal factor but instead itself an effect of the Russian social environment.[11]

Such approaches to the study of the Russian Revolution have certainly advanced our understanding of the period. Yet with few exceptions, revisionist accounts of the fall of tsarism and of the events of 1917 focus their attention on the events leading up to the Bolshevik seizure of power, rather than examine in detail the subsequent process of political institution building under Bolshevik auspices. The implicit assumption of much of this work seems to be that if the motive force for the destruction of the old order can be traced to broader social groups, then the design of the formal political system emerging in the Soviet period must also reflect the interests of these groups more than the ideologies of revolutionary elites.

Logically, however, this conclusion is by no means necessary. Recently, theorists of revolution have begun to emphasize that even though committed ideological groups may play only a minor role in creating revolutionary situations, the ideology of victorious revolutionary elites may nonetheless have a crucial effect on the development of postrevolutionary institutions. The term *revolution* itself, from this point of view, tends to conflate two very different processes—the breakdown of existing social structures and the formation of new ones—whose dynamics must be studied separately.[12] In the Russian case, then, Lenin's ideology and his concepts of political and social organization may have had very little to do with the collapse of tsarism, or even of the Provisional Government; they were nonetheless critical in shaping Soviet political institutions after the Bolshevik takeover.

It is in this context that I propose to examine the links between Marx's charismatic-rational conception of time and Lenin's design of the central political institution of the Soviet state from 1917 to 1991—the Communist Party. In this chapter I argue that Lenin's conception of the party represents an ingenious, but by no means inevitable, political solution to the paradox of time and political action inherent in the Marxist theoretical framework. After a detailed examination of Lenin's major treatise on political organization, *What Is to Be Done?*, I address the question of how Lenin proposed to institutionalize socialism after the revolutionary conquest of political power, drawing primarily on his crucial but misunderstood work *The State and Revolution.* Finally, I examine the contradictions implicit in Lenin's postrevolutionary discussions of the use of time in Soviet economic policy, arguing that while Lenin successfully welded together aspects of rational and charismatic time in the political practice of the Bolshevik Party, these elements began to fall apart again once Lenin turned to the question of how to organize economic institutions and to inculcate a culture of revolutionary discipline.

As in the Second International, I argue, these contradictions at the economic and cultural levels resulted in a new split in the (now) Marxist-Leninist movement. Trotsky's "left," or charismatic, Leninism, Bukharin's "right," or rational-legal, Leninism, and Zinoviev's "orthodox," or neotraditional, Leninism developed in ways that closely paralleled the earlier political debates between Luxemburg, Bernstein, and Kautsky. The resulting factionalization of the party created a political crisis that allowed a new innovator, Stalin, successfully to impose a new synthesis of time transcendence and time discipline in economic activity—just as Lenin's politics had emerged out of the breakdown of the Second International during World War I. Thus, the period between Lenin's political takeover as leader of the Bolshevik revolution in 1917 and the end of the New Economic Policy (NEP) period in the 1920s can be considered as a second cycle of development in the history of the Marxist charismatic-rational conception of time—the political cycle.

What Is to Be Done?

The key elements of Lenin's successful institutionalization of the charismatic-rational conception of time on the political level are already contained in his famous treatise on party organization, *What Is to Be Done?* Written in 1901–2 as a polemic against the influence of Bernstein's revisionism on Russian Social Democracy, *What Is to Be Done?* is remarkable both for its incisive analysis of right and left tendencies within the revolutionary movement and for its detailed institutional response to the apparent inadequacies of both from a Marxist perspective. Indeed, it could be argued that as far as questions of political organization are concerned, Lenin never really added anything to what he wrote in his first major theoretical work. In his later treatises on the split with the Mensheviks in 1903, on the collapse of Kautsky's SPD in 1914, and on the rise of "left-wing communism" in Germany and Russia after 1917, he rarely needed to go beyond the principles he had formulated years before.

However, a full understanding of *What Is to Be Done?* is impossible without placing this work in its historical context.[13] To recapitulate, the dilemma faced by the Marxists of the Second International after the deaths of Marx and Engels was as follows: the revisionism of Bernstein and his followers necessarily sacrificed the charismatic transcendence of time promised by Marx in favor of a time-bound, evolutionary process of amelioration of the capitalist system, while attempts at revolutionary action here and now, as favored by Rosa Luxemburg, disconnected political practice from the analysis of the temporal laws of history that alone could determine when the time for revolution would be ripe. The left was thus subject to the danger of revolutionary failure, the right to the danger of purely bourgeois success. How, then, could the need for revolutionary action and the need for respecting the impersonal force of time be reconciled politically?

These issues were, if anything, even more troubling for Russian Marxists than they were for the Germans. Plekhanov, who had founded Russian Social Democracy in 1883, had done his best to establish the movement on a thoroughly orthodox basis, echoing the writings of Engels and Kautsky as consistently as possible, given Russian conditions.[14] The line of the Emancipation of Labor Group, the main theoretical center of Russian Marxism in exile, was essentially the same as that of the Kautskyan "center" discussed previously: fidelity to the Marxist "dialectical method" of analysis would be rewarded with "inevitable" revolutionary success; the main thing was to avoid deviations from scientific socialism. But if in Germany this orthodox analysis tended to lead to a sort of "revolutionary waiting," in the extremely undeveloped economic conditions of tsarist Russia it would appear to mandate absolute passivity—despite Plekhanov's revolutionary phraseology. Marxists in "feudal" Russia would have to wait not only for socialism but for capitalism as well. Radical workers in Moscow and Saint Petersburg, who were already actively protesting for shorter working hours and political freedoms, would have to be told that any attempt by the proletariat to lead a revolution in the foreseeable future was "premature." In such an environment, the temptation for Marxists to adopt a "right" reformist or "left" revolutionary position was perhaps even greater than in developed western Europe. Helping the workers struggle for better working conditions here and now, or engaging in direct revolutionary action against the tsarist autocracy, at least allowed those who opposed the status quo to *do* something—even if neither of these paths was fully consistent with the Marxist synthesis of charisma and rationality.

At the same time, strict adherence to Marxist theory allowed one to predict the ultimate failure of pure reformism and pure revolutionism alike in the autocratic Russian social environment. Under predominantly feudal conditions, reformists without a truly revolutionary vision would be betrayed by the weak Russian bourgeoisie—itself hopelessly dependent upon the tsarist state—while revolutionists without rational discipline would quickly be rounded up, arrested, and possibly executed. Indeed, the experience of Russian radicalism during the nineteenth century had amply demonstrated the bankruptcy of both left and right approaches.

Considerations such as these led the young Lenin to adopt a fundamentally orthodox Marxism, siding with Plekhanov against both revisionist and terrorist tendencies within the emerging Russian Marxist movement. In the broader context of the Second International, Lenin considered himself an ally of Kautsky as well. However, Lenin also sensed the danger of orthodoxy's degeneration into passivity in a way neither Plekhanov nor Kautsky quite grasped—as is evident from the confusion of the two elder Marxists when they later found themselves labeled "traitors" and "renegades" by the man each had previously considered a staunch ally. Indeed, by 1902, Lenin had already developed an independent position within the context of Russian Marxism that marked him as a institutional innovator within the context of the charismatic-rational conception of time.

What Is to Be Done? was written as an attack on Bernstein's adherents in Russia, the so-called Economists, who emphasized the importance of trade union struggles and economic demands over direct political struggle with the tsarist autocracy. However, while the focus of the work is thus logically on the failings of the Economists' interpretation of socialist strategy, Lenin expands his attack to include a condemnation of the most significant left-wing alternative to Economism then existing in Russia, the terrorism of the peasant-oriented Socialist Revolutionary Party. Although *What Is to Be Done?* preceded by a few years the full-fledged articulation in Germany and elsewhere of Luxemburg's type of left Marxism, Lenin's arguments against premature revolutionary action in this work differ little from those he would advance in later polemics against leftism.

The first element of Lenin's solution to the dilemma of time in political Marxism was largely taken from Kautsky and Plekhanov: a call for socialist intellectuals to combat "spontaneity." However, the problem of spontaneity receives much greater theoretical attention in Lenin's work than in that of the previous guardians of orthodoxy. According to Lenin, a "slavish cringing before spontaneity" was in fact common to *both* terrorism and Economism (and by extension, to both left and right Marxism).[15] The only difference between the two, Lenin argued, was that they "bow[ed] to opposite poles of spontaneity"; that is, Economists bowed to the spontaneity of "the labor movement pure and simple," whereas terrorists bowed to the spontaneity of "intellectuals, who lack the ability or opportunity to connect the revolutionary struggle and the working-class movement into an integral whole."[16]

The problem with spontaneity, from Lenin's orthodox Marxist perspective, was that in order to make human will an effective force, the material and historical forces shaping one's social milieu had first to be understood and exposed. As Marx put it, "Men make their own history, but not of their own free will, not under circumstances that they themselves have chosen."[17] Any attempt to "make history" without an understanding of the constraints placed on human action by the force of time was doomed to failure. Spontaneity, as opposed to the historically grounded will of Marxist revolutionaries, could thus never be a force for revolutionary change; rather, by rejecting the need to combine revolutionary will with discipline, those who favored spontaneous action by workers or by terrorist intellectuals could only solidify the hold of the dominant bourgeois social order. Spontaneous action by the proletariat could lead only to the growth of "trade union consciousness" among the workers; spontaneous action by terrorists could only fragment the socialist movement, making it more susceptible to police raids.[18]

In essence, Lenin recognized that to rely on any sort of spontaneity as the main force for bringing about socialism was to give in to the power of time, to become a slave to existing circumstances—in short, to cease being a revolutionary. Spontaneous action gave only the illusion of free will; in fact, a reliance on spontaneity unavoidably underestimated the power of the temporal historical forces constraining human action and therefore amounted to a de facto surrender to those forces: "We revolutionary democrats . . . are dissatisfied with this worship of spontaneity, i.e., of that which exists 'at the present moment.' . . . In a word, the Germans stand for that which exists and reject changes; we demand a change of that which exists, and reject subservience thereto and reconciliation to it."[19] But as Lenin recognized, understanding the fundamental similarity of left and right interpretations of Marxism counted for little in the absence of an effective institutional alternative to a reliance on trade unions or terrorist cells. As he himself put it, "If you have no organizational ideas *of your own*, then all your exertions in behalf of the 'masses' and 'average people' will be simply boring."[20]

In this light, Lenin's proposal to transform the diffuse Russian Social Democratic movement into a disciplined, hierarchical party of "professional revolutionaries" was a brilliant solution to the dilemma of political institutionalization facing the Marxist cause.[21] The very phrase "professional revolutionary" neatly combines the charismatic essence of revolutionary action with the rational time discipline characteristic of the modern professional, expressing succinctly both sides of the charismatic-rational conception of time in organizational terms. In this sense, a party of professional revolutionaries could legitimately claim to be faithful to the Marxist legacy, despite the lack of any explicit discussion of the need for such a centralized organization in Marx's work.

If the spontaneity of intellectuals led to a reliance on charismatic forms of revolutionary activity devoid of rational time discipline, while the spontaneity of workers led to an essentially evolutionary process of reform devoid of revolutionary content, Lenin's party of professional revolutionaries had to include institutional guarantees against both forms of spontaneity. Lenin's answer to this problem was to stress two key components of party organization: theory and discipline. Like Kautsky, Lenin argued that fidelity to revolutionary theory would be the guarantor of the movement's socialist purity. Going beyond Kautsky, Lenin insisted that the Marxist theoretical synthesis of time discipline and time transcendence should become the basis of a "strict party discipline" governing every member's daily use of time.

Theory—that is, Marxist theory as interpreted by the party leadership—would ensure that the spontaneous workers' movement would continue to be directed toward revolutionary goals, despite the necessarily gradual character of the struggle for economic concessions. As Kautsky had correctly argued, the revolutionary aims embodied in Marxist theory could not be achieved by the workers on their own, since Marxism itself, the sole possible strategic foundation for correct revolutionary tactics, was formulated by intellectuals who "themselves belonged to the bourgeois intelligentsia."[22] To subordinate revolutionary activity to the "purely economic demands" of the workers, as the Economists advocated, was thus to abdicate the intellectuals' responsibility for ensuring the theoretical purity and revolutionary essence of the movement. As Lenin concluded, "without revolutionary theory, there can be no revolutionary movement."[23]

However, although Lenin's stress on revolutionary theory was essentially Kautskyan, his conception of party discipline represented a wholly original view of how to institutionalize charismatic-rational time among revolutionary socialists. Revolutionary time discipline in the Leninist party encompassed both external party strategy and internal party activity. In terms of external strategy, party discipline would ensure that the timing of revolutionary attacks would be well planned. Such a sense of timing is a necessary component of revolutionary communism, Lenin insists; indeed, the workers' increasing ability to carry out "carefully timed" strikes is a sign of their growing class consciousness.[24] But, he argues, it takes a strictly centralized party organization to ensure that all revolutionary offensives and retreats are conducted in a timely manner. "A strong revolutionary organization is absolutely necessary," Lenin writes, "precisely for the purpose of giving stability to the movement."[25] Unlike "the utterly unsound Economism and the preaching of moderation, and the equally unsound 'excitative terror,' which strives 'artificially to call forth symptoms of the end of the movement . . . when this movement is as yet nearer to the start than the end,'" Lenin's disciplined party would "safeguard the movement against making thoughtless attacks and prepare attacks that hold out the promise of success."[26]

Internally, "professional" party discipline would overcome the sporadic nature of revolutionary activity due to the periodic raids of the police and to the movement's overreliance on mass spontaneous action. The party, Lenin argued, must "consist chiefly of people professionally engaged in revolutionary activity," since "in an autocratic state, the more we *confine* the membership of such an organization to people who are professionally engaged in revolutionary activity and who have been professionally trained in the art of combating the political police, the more difficult it will be to unearth the organization."[27] Of course, Lenin by no means excludes the possibility of working with more informal organizations, such as trade unions, workers' study circles, and so on. He insists, however,

> It would be absurd and harmful *to confound* them with the organization of *revolutionaries*, to efface the border-line between them, to make still more hazy the all too faint recognition of the fact that in order to "serve" the mass movement we must have people who will devote themselves exclusively to Social-Democratic activities, and that such people must *train* themselves patiently and steadfastly to be professional revolutionaries.[28]

By the same token, workers promoted to party membership must be given the financial means to free themselves from bourgeois factory discipline in order to adopt truly professional party discipline: "A worker-agitator who is at all gifted and 'promising' *must not be left* to work eleven hours a day in a factory. We must arrange that he be maintained by the Party; that he may go underground in good time; that he change the place of his activity, if he is to enlarge his experience, widen his outlook, and be able to hold out for at least a few years in the struggle against the gendarmes."[29] In essence, Lenin's conception of party discipline demanded that the party member subordinate his personal desire to transcend time to the strict revolutionary time discipline of the party. The party itself, then, became the collective charismatic agent that would master time within the boundaries of time.

Lenin's "party of professional revolutionaries" thus *preserved the ethos* of Marxism while addressing the strategic problems of revolutionary politics. As we have seen, no faction within the dominant socialist party of the Second International, the German SPD, successfully fused the charismatic and rational elements of Marx's view of time while providing a workable response to the problem of Marxism's political institutionalization. Lenin's "party of a new type," with its stress on maintaining orthodox theory as the guarantor of the party's revolutionary essence and with its enforcement of discipline as the guarantor of its professionalism and effectiveness, did both. By making the party, rather than the intellectuals or the proletariat, the locus of revolutionary charisma, by identifying it as the sole possible agent of time transcendence, Lenin could demand a high degree of essentially rational time discipline *within* the party without sacrificing the members' sense that they were contributing to the destruction of bourgeois time.

Indeed, Lenin's party, as he himself intuitively grasped, could be seen as a sort of communism in embryo at work in the present, empirical world. Since within the party all were subject to the same revolutionary discipline, "all *distinctions between workers and intellectuals*, not to speak of distinctions of trade and profession, in both categories, must *be effaced*"—a situation that closely resembled Marx's vision of the overcoming of the distinction between manual and mental labor under communism.[30] In addition, such an organization would internally overcome the alienation of the individual from the collective inherent in class society. As Lenin put it in arguing against the principle of procedurally democratic elections within the party, the achievement of a secret, strictly professional revolutionary organization would guarantee "something more than 'democratism' . . . namely, complete, comradely, mutual confidence among revolutionaries."[31] By submitting oneself to party discipline, then, one could be *both* more effective than Economists or terrorists, than Bernsteinian or Luxemburgian Marxists, *and* participate more surely in a social order akin to the original vision of communism that inspired the Marxist movement.

Leninism before 1917

With *What Is to Be Done?*, Lenin theoretically transcended the struggle between the right, left, and center factions in the Second International—a struggle that nonetheless continued for a decade afterward. Once he found an institutional vehicle for combating "right spontaneity" and "left spontaneity" alike, his later works on party organization more or less consistently expanded on his original arguments, attacking "revisionist" opponents such as the Mensheviks, as well as left-wing communists both in Russia and abroad, in increasingly polemical terms.[32] Lenin's legendary genius for tactical maneuvering vis-à-vis his opponents in the socialist camp—his ability to cast aside close comrades who had drifted from him theoretically, as well as his ability to co-opt those who had once disagreed with him but later became useful to the Bolshevik cause—should be understood as the result of his simultaneous theoretical rejection of bourgeois accommodation and revolutionary romanticism, the two trends that had seduced practically every other Marxist theorist of the period, rather than as an unprincipled series of purely instrumental political moves. Having reconciled the contradictions of left and right in the political institution of the Bolshevik Party, Lenin could side with either group on particular issues without sacrificing his own theoretical synthesis—and without succumbing to an orthodox "revolutionary" passivity.

However, to emphasize Lenin's theoretical consistency in defending the institutional design of his "party of a new type" does not imply that the political practice of the actual Bolshevik Party from 1902 to 1917 neatly conformed to Lenin's model of professional revolutionary discipline. If anything, the uncompromising centralism Lenin advocated tended to alienate even those who in principle were supportive of Lenin's staunchly revolutionary position. Moreover, to expect that a disciplined, conspiratorial party of professional revolutionaries would function smoothly in the turbulent and repressive environment of tsarist Russia was, of course, unrealistic. As Robert Williams and others have shown, the prerevolutionary Bolshevik Party was in fact a highly factionalized, theoretically diverse, and sometimes practically defunct association that certainly bore little or no resemblance to the model set out in *What Is to Be Done?*[33]

It would, nonetheless, be a profound mistake to conclude from this that Lenin's early conception of the party played no significant role in orienting later Bolshevik or Soviet politics. Indeed, given the obstacles mentioned above, it is all the more remarkable that throughout his political life, Lenin kept trying to realize his vision of charismatic-rational political discipline—even when this threatened to eliminate the empirically existing party entirely! And after the Bolshevik seizure of power, once it became clear that the world proletarian revolution would not immediately succeed, Lenin began to forge a party dictatorship that eventually approached the formal ideal type he had set out in his 1902 treatise.

The Bolshevik Party itself originally emerged as a result of Lenin's insistence at the Second Congress of Russian Social Democrats in 1903 that those unwilling to work as full-time "professional revolutionaries" be excluded from membership in the movement. Lenin's intransigence on this issue—the doctrinal importance of which can only be understood in terms of Lenin's argument in *What Is to Be Done?*—cost him the support of erstwhile allies such as Martov, Akselrod, and Zasulich. Even though they did not all disagree with him on the technical point in question, they failed to understand why Lenin saw exclusion of part-time socialists as the defining test of fidelity to Marxism. By 1904, Lenin's uncompromising stance on the need to institutionalize party discipline alienated Plekhanov as well, who rejoined his former colleagues in the Menshevik group and began to castigate Lenin's dogmatism and "dictatorial tendencies." Henceforth Lenin consistently interpreted all Menshevik criticisms of his doctrine as de facto capitulations to Bernsteinian revisionism.

Forced to search for new supporters, Lenin began to collaborate with a group of rather mystical Marxists centered around the philosopher and science fiction writer Aleksandr Bogdanov, who were attracted to what they saw as Lenin's pure form of revolutionism in comparison with that of the Mensheviks. But they, too, soon began to deviate from Lenin's definition of theoretical and political orthodoxy—this time in a leftist direction. If Plekhanov and the other Mensheviks appeared to counsel a passive embrace of development within linear, abstract time, Bogdanov's left Bolsheviks slid into a purely charismatic, undisciplined irrationalism—most spectacularly illustrated in Anatolii Lunacharsky's project to turn socialism into a new "religion" for the masses. The Bogdanovites' attempt to train young Russian socialists at a "Bolshevik" school at Maksim Gorky's villa on the island of Capri led Lenin to break with them as well—even though this left him, by 1912, almost completely isolated within the Russian Social Democratic movement. However, the few disciples who remained loyal to Lenin throughout the 1912–17 period—notably Zinoviev, Kamenev, and Stalin—were later rewarded with powerful positions in the Bolshevik Party elite.

Interestingly, Lenin's lengthy philosophical treatise arguing against the quasi-religious components of Bogdanov's Marxism, *Materialism and Empirio-Criticism*, also contained a detailed examination of the problem of time. Since this work became the ritualistic starting point for all later Soviet investigations of the problem of time in philosophy, it is worth summarizing its main themes. Closely following Engels's analysis set out in the *Anti-Dühring*, Lenin maintains that nothing can exist outside time or space, and accuses those who assert the contrary of betraying "materialism" in favor of "idealism": "It is impossible to hold consistently to a standpoint in philosophy which is inimical to all forms of fideism and idealism if we do not definitely and resolutely recognize that our developing notions of time and space *reflect* an objectively real time and space."[34] The Bogdanovites, by embracing the views of Ernest Mach on the subjectivity and relativity of time and space, had "opened the door" to all sorts of speculative and reactionary tendencies within the socialist movement. In short, they had substituted the religious form of charisma outside time for the Marxist synthesis of human time transcendence within time:

> Like all the Machians, [they] erred in confounding the mutability of human conceptions of time and space, their exclusively relative character, with the immutability of the fact that man and nature exist only in time and space, and that beings outside time and space, as invented by the priests and maintained by the imagination of the ignorant and downtrodden mass of humanity, are disordered fantasies, the artifices of philosophical idealism—useless products of a useless social system.[35]

The denial of time constraints by beings who necessarily live within time thus meant an abandonment of rationality and science.

At the same time, however, Lenin was keen to show that accepting the inexorability of "real" time does not prevent effective action to transform reality. Again, he took his cue from Engels, expanding the latter's definition of freedom as the appreciation of necessity into a call for action as the ultimate test of truth:

> Engels plainly employs the *salto vitale* method in philosophy, that is to say, he makes a *leap* from theory to practice. Not a single one of the learned (and stupid) professors of philosophy would permit himself to make such a leap, for this would be a disgraceful thing for a devotee of "pure science" to do. For them the theory of knowledge . . . is one thing, while practice is another. For Engels all living human practice permeates the theory of knowledge itself and provides an *objective* criterion of truth. For until we know a law of nature, it, existing and acting independently and outside our mind, makes us slaves of "blind necessity." But once we come to know this law, which acts (as Marx pointed out a thousand times) *independently* of our will and our mind, we become the lords of nature.[36]

Here we see the charismatic-rational conception of time expressed with utmost clarity. Despite all of his excoriation of the leftists, like Bogdanov and Lunacharsky, who denied the scientific reality of time, Lenin saw those who refused to "leap" from scientific theory to revolutionary practice, like the Bernsteinians and the Mensheviks, as equally reactionary. Since, according to Lenin, Marx had discovered "laws" of social development analogous to the natural laws of Newtonian physics and Darwinian evolutionary biology, Lenin's position amounted to a claim that proper mastery of Marxism would put humanity as a whole in control of all of nature, even while remaining within time and space.[37]

Not long after the publication of *Materialism and Empirio-Criticism*, in a 1913 encyclopedia entry on Marxism, Lenin presented perhaps the fullest explanation of how the charismatic-rational synthesis of time transcendence and time discipline could be translated into the political strategy of proletarian revolution. In the context of the argument so far, it is worth quoting at some length:

> Only an objective consideration of the sum total of the relations between absolutely all the classes in a given society, and consequently a consideration of the objective stage of development reached by that society and of the relations between it and other societies, can serve as a basis for the correct tactics of an advanced class. At the same time, all classes and all countries are regarded, not statically, but dynamically, i.e., not in a state of immobility, but in motion. . . . Motion, in its turn, is regarded from the standpoint, not only of the past, but also of

the future, and that not in the vulgar sense it is understood in by the "evolutionists," who see only slow changes, but dialectically: ". . . in developments of such magnitude twenty years are no more than a day," Marx wrote to Engels, "though later on there may come days in which twenty years are embodied." At each stage of development, at each moment, proletarian tactics must take account of this objectively inevitable dialectics of human history, on the one hand, utilizing the periods of political stagnation or of sluggish, so-called "peaceful" development in order to develop the class-consciousness, strength and militancy of the advanced class, and, on the other hand, directing all the work of this utilization towards the "ultimate aim" of that class's advance, towards creating in it the ability to find practical solutions for great tasks in the great days, in which "twenty years are embodied."[38]

An understanding of Marx's metaphor of a single day "embodying" twenty years, and vice versa, is here put forth as the very foundation of successful socialist strategy. For Lenin, time was indeed absolute, as Newton and Engels had insisted. All who believed in pure time transcendence were "idealists" who would never be able to translate their beliefs into time-bound reality. But departing from the Newtonian conception of time as an abstract grid, Lenin argued that the "content" of time—the qualitative development of socialist "consciousness" within it—continually varied.

From the purely charismatic point of view, time is an illusion that can be miraculously transcended by extraordinary individuals. From the purely rational point of view, each period of one day or twenty years is *temporally* exactly the same as any other quantitatively equivalent period. Lenin, by contrast, argued that proper proletarian tactics demanded an understanding of time as flexible: relatively elongated in periods of reaction, relatively compressed in periods of revolutionary advance. The role of the party of professional revolutionaries was thus to gauge what "sort of time" it was and direct the proletarian movement accordingly. What had been only hinted at in Marx's theoretical work thus became the basis for formal institutional behavior in Lenin's politics: the realization of socialism as a form of charismatic-rational time discipline.

Time in *Imperialism* and *The State and Revolution*

Two crucial questions concerning the relationship of Marxist theory to revolutionary politics remained unaddressed in *What Is to Be Done?*

and in Lenin's writings before 1914. First, the idea that the "professional revolutionary" party could lead a proletarian revolution in a backward, peasant country such as Russia, without waiting for the proletariat of the more advanced capitalist countries to begin the process, was never made explicit in Lenin's early work. Second, Lenin had not yet confronted the question of how a victorious "party of professional revolutionaries" coming to power in one country should begin to organize socioeconomic activity. Lenin's views on these two questions can be investigated through an examination of the two major essays he wrote between the collapse of the Second International in 1914 and the Bolshevik revolution of October 1917: *Imperialism* and *The State and Revolution.*

Imperialism marks a crucial stage in Lenin's emergence as a revolutionary political innovator rather than a mere defender of Marxist orthodoxy. Specifically, in this work Lenin breaks with the third tendency within the Second International described in the previous chapter—the centrism of Karl Kautsky. Because Kautsky had always been a staunch opponent of both Bernstein and Luxemburg, Lenin had remained in close theoretical agreement with him throughout the prewar years. Kautsky in this period preached revolutionary purity to the Marxist movement as a whole; Lenin took what Kautsky said about international Marxism seriously and applied it to the Social Democratic movement in Russia.

Thus it came as a huge shock to Lenin when Kautsky, along with almost the entire leadership of the Second International, refused to struggle against the participation of the bourgeois states of Europe in World War I—indeed, even abstaining while the majority of the SPD voted in parliament for the extension of war credits to the "imperialists." How could someone who seemed to understand the core principles of Marxism so deeply make such a treasonous concession to the bourgeois status quo? In fact, Kautsky was not at heart a revolutionary but an orthodox "centrist" whose policy was one of "revolutionary waiting" rather than "revolution making." However, Lenin did not fully realize, until 1914, the nature of the difference between his form of orthodoxy and Kautsky's.

Now, in *Imperialism,* Lenin accomplished three things simultaneously: he debunked Kautsky's theories of international capitalism as representing a new form of center revisionism in the Marxist movement, he set out his own alternative theory showing why the time was ripe for the global overthrow of capitalism, and finally he demonstrated why proletarian revolution was *unlikely* to begin in the core of capitalism without a prior revolutionary success in the periphery. As a result, *Imperialism* provided Lenin with the theoretical grounding for an attempt to launch proletarian revolution *now*, in peasant Russia, rather than waiting for the maturation of the "productive forces."[39]

Kautsky's essential departure from Marxism, Lenin argued, lay in his assumption that imperialism and monopoly capitalism might create the possibility of a coordinated form of exploitation compatible with "peaceful" relations between capitalist states. Imperialism, Kautsky argued, might evolve through a series of intercapitalist alliances into a sort of "super-imperialism" based on the joint rule of the world bourgeoisie. This "super-imperialism" might prove to be extremely stable, thus blunting the immediate prospects for proletarian revolution. Lenin dismissed Kautsky's argument:

> We ask, is it "conceivable," assuming that the capitalist system remains intact—and this is precisely the assumption that Kautsky does make—that such alliances would be more than temporary, that they would eliminate friction, conflicts and struggle in every possible form?
>
> The question has only to be presented clearly for any other than a negative answer to be impossible.[40]

In fact, Lenin continued, alliances between capitalist states are never more than a sort of " 'truce' in periods between wars." Indeed, it is the fundamental instability of these alliances that gives the Marxist revolutionary movement some hope of success; to gloss over such intercapitalist tensions, as Kautsky did, can only be an excuse for inaction:

> The only objective, i.e., real, social significance of Kautsky's "theory" is this: it is a most reactionary method of consoling the masses with hopes of permanent peace being possible under capitalism, by distracting their attention from the sharp antagonisms and acute problems of the present times, and directing it towards illusory prospects of an imaginary "ultra-imperialism" of the future. Deception of the masses—that is all there is in Kautsky's "Marxist" theory.[41]

Lenin's own argument was different: imperialism, based on financial and commercial monopolies that had extended the rule of capital throughout the entire world, was the "highest stage of capitalism." But this form of monopoly capitalism, like the laissez-faire economic system that had given rise to it, was subject to severe contradictions that would work to undermine its stability. Competition between imperialist states inevitably produced wars that undermined the solidarity of capitalists against the world proletariat. Furthermore, precisely because imperialism had pushed capitalism beyond narrow national boundaries and made it a world system, the question of whether individual countries were "ripe" for revolution had become anachronistic. If imperialism was the "highest stage of capitalism," then the next "stage" of world development could only be socialism. Essentially, Lenin's argument implied that a successful revolutionary overthrow of bourgeois rule anywhere in the world would bring down the whole structure of international capitalism.

But the argument of *Imperialism* went even further in attacking Kautsky's orthodox interpretation of Marxism, for Lenin here began to make sense of the "opportunism" of the Second International as *itself* a product of imperialism! In the colonial powers, Lenin argued, the super-profits derived from imperialist exploitation tend to create "privileged sections . . . among the workers" who are "detach[ed] . . . from the broad masses of the proletariat." As a result, in advanced capitalist states, imperialism has a tendency "to split the workers, to strengthen opportunism among them and to cause temporary decay in the working-class movement."[42] Lenin here subtly unveiled a materialist explanation for the decline of the Second International as a revolutionary force: it was an expression, within the Marxist movement itself, of this very process of "temporary decay." The implicit message, once again, was that the renewal of the revolutionary energies of the world proletariat would have to begin where the class struggle was still carried out in a "purer" form—that is, where all forms of "opportunism" in the Marxist movement were excluded by a political organization capable of ensuring true "revolutionary discipline."

The publication of *Imperialism* marked Lenin's final break with Kautskyan orthodoxy and his assertion of the indispensability of revolutionary *action* over "revolutionary waiting." But the second theoretical problem that faced Lenin's project of party-led revolution in Russia— the question of how a victorious vanguard party might organize economic production in a "proletarian state," especially one still dominated by peasant agricultural production—proved to be more intractable. Lenin, who often despaired of ever living to see a victorious proletarian revolution in Russia, left this question unaddressed until only a few months before the Bolshevik revolution itself. As if suddenly made aware of the importance of what he had left unsaid in his earlier works as the proletarian revolution actually approached, Lenin spent August and part of September 1917, while in hiding from the Provisional Government, making the final revisions to another highly significant theoretical work: *The State and Revolution*.[43]

The State and Revolution, as A. J. Polan has pointed out, has often been treated merely as an uncharacteristic outburst of Lenin's residual Marxist "Utopianism."[44] The assumption here is that since the Utopian ideals expressed in the work—such as the performance of all state administration by the working masses and the concomitant gradual disappearance of the state itself under socialism—were never actually implemented in the Soviet regime, *The State and Revolution* can be safely ignored and we should focus instead only on Lenin's "practical" political works, such as *What Is to Be Done?* Not only does such an artificial division between Lenin's writings rely on a misleading dichotomy between "ideology" and "pragmatism" in his writings—as if his practical politics were not informed by his theoretical worldview—but it ignores the integral and important connection between Lenin's views on party organization and his views on state and economic organization. Real differences of emphasis and argumentation between the two works exist, of course—but not because Lenin unaccountably indulged himself in Utopianism just before returning to pragmatic revolutionary activity. Rather, in *The State and Revolution* Lenin simply addressed himself for the first time to a question he had previously thought it was premature to discuss: how to organize a postrevolutionary socialist society.

At the same time, one cannot help but be struck by the fundamentally unpersuasive nature of Lenin's argument on this topic, especially as compared with the ruthless logical flow of *What Is to Be Done?* Lenin, seemingly feeling his way, relies much more heavily on direct quotes from Marx and Engels in this work than in previous writings, and the overall organization of the essay is less sure. However, the unpersuasive quality of *The State and Revolution* is not due to Lenin's misplaced idealism but rather to the fact that in addressing the question of the postrevolutionary state, Lenin unsuspectingly stumbled into the second unresolved dichotomy between charisma and rationality in Marx's work—that underlying Marx's conception of the use of time in socioeconomic institutions under socialism. Having ingeniously resolved the dilemma of time discipline versus time transcendence in the sphere of political action, Lenin was unable to find a similarly effective institutional response to the problem of how to create socioeconomic institutions that would be simultaneously "revolutionary" and "professional."

The problem was also evident in Lenin's views on the related debate over the role of the state in guiding economic policy in the period after a successful political revolution. Should the Bolsheviks work to destroy the state and let socialist economics emerge spontaneously? Or should the state try to enforce socialist economic norms on a primarily peasant population? Ultimately, rather than resolving this dilemma, Lenin tried to straddle both sides of it simultaneously.

The question of state power after the proletarian revolution, like most issues connected with the actual organization of socialist society, was addressed only tangentially in the work of Marx and Engels. As we have seen, Marx dismissed the Hegelian notion that the state could become a force for the reconciliation of abstract morality and concrete social identity, and his early philosophical battles with more statist Hegelians left a strong antistatist imprint on his writings on this issue. Fundamentally, Marx saw the bourgeois state, like all state organization in class society, as essentially a vehicle for the domination of the ruling class; therefore, logically, after the proletarian revolution state organization as such would eventually disappear—along with classes themselves.

The question begged here is, of course, precisely *when* the disappearance of the state might take place. But keeping in mind Marx's charismatic-rational conception of time, this question takes on a special form that is otherwise likely to be missed. From this perspective, the question cannot be answered either purely "charismatically," by asserting that the state will disappear immediately and absolutely as soon as the revolution is victorious, or according to a purely rational system of time measurement, that is, by specifying some more or less accurate interval of abstract time before the abolition of the state—in years, decades, or centuries. The first answer descends into anarchism and ignores the empirical continuities between pre- and postrevolutionary society that must be acknowledged if one adopts a scientific, rational viewpoint on the nature of the unfolding of events in time. The second answer, however, completely denies the transformative essence of revolution, which marks off the postrevolutionary era as a *qualitatively different* time period from what went before. But if neither of these responses is correct, it is not at all clear what positive program a "socialist state" should adopt.

The bulk of Lenin's *The State and Revolution* finds itself enmeshed in these very contradictions. Following Marx and Engels, Lenin insists that the proletariat needs to utilize state power for an indefinite period in order to fight the continuing resistance of the capitalists: "The proletariat needs the state only temporarily. We do not at all differ with the anarchists on the question of the abolition of the state as the *aim*. We maintain that, to achieve this aim, we must temporarily make use of the instruments, resources and methods of state power *against* the exploiters, just as the temporary dictatorship of the oppressed class is necessary for the abolition of classes."[45] The problem becomes, given the charismatic-rational conception of time, just how does one interpret the word "temporary" here? Is the proletarian state merely the last vestige of bourgeois time's existence, destined to disappear along with time constraints themselves almost immediately after revolutionary victory? Or is the proletarian state merely the bourgeois state with a new leadership, the enforcer of socioeconomic time discipline under new auspices? Lenin tries to give both answers at once: the postrevolutionary state must be *both* proletarian and bourgeois, based simultaneously on a faith that time constraints will be overcome and on a recognition of the need for societywide enforcement of time discipline.[46]

This dichotomy in Lenin's work can be illustrated by contrasting two quite different sections of *The State and Revolution*, each of which expresses one side of the charisma/rationality dualism. The first deals with the need for the revolutionary proletariat to "smash" the bourgeois state machine and outlines the form of the proletarian "semi-state" that will replace it, based on Marx's analysis of the Paris Commune of 1871.[47] The second deals with questions of labor discipline in the "protracted" interim period between socialism and communism.

The first section, which includes the first four chapters of the work, clearly emphasizes the charismatic, transformative nature of revolution and assumes that bourgeois time constraints in the postrevolutionary era have already been overcome to a significant degree. The institutional outline of the proletarian semistate—which is, according to Engels, "no longer a state in the proper sense of the word"—clearly reflects this assumption of practical timelessness.[48] The commune-state, in Marx's words, is to be "a working, not a parliamentary body"; such a state will not lose its organic unity with the masses, Lenin argues, since "the parliamentarians themselves have to work, have to execute their own laws, have themselves to test the results achieved in reality, and to account directly to their constituents."[49] For an individual adequately to perform such a wide-ranging list of responsibilities would require a practically infinite amount of time; as Polan has put it, "The only question seems to be of what such an individual would die: overwork or multiple schizophrenia."[50] But such problems are not germane to Lenin's analysis here, operating as he is on the assumption of revolutionary transcendence of ordinary time constraints.

The same denial of time constraints enters into Lenin's description of economic organization under socialism, under which each and every member of the proletariat eventually takes part in basic administrative work:

> *We*, the workers, shall organize large-scale production on the basis of what capitalism has already created, relying on our own experience as workers, establishing strict, iron discipline backed up by the state power of the armed workers. We shall reduce the role of state officials to that of simply carrying out our instructions as responsible, revocable, modestly paid "foremen and accountants." . . . Such a beginning, on the basis of large-scale production, will of itself lead to the gradual "withering away" of all bureaucracy, to the gradual creation of an order . . . under which the functions of control and accounting, becoming more and more simple, will be performed by each in turn, will then become a habit and will finally die out as the *special* functions of a special section of the population.[51]

To be sure, Lenin's insistence on the need for "strict, iron discipline" remains here, but with two fundamental caveats. First of all, it will apparently take no time at all to replace irresponsible or lazy officials under socialism, since their positions will be immediately "revocable"; time constraints on the training of new officials apparently do not present any problem. Second, the state of "armed workers" that enforces socialist time discipline soon ceases to be a state at all, and the gradually increasing participation of everyone in "control and accounting" transforms external discipline into a sort of spontaneous self-discipline. The question of whether or not, under conditions of complete freedom, all workers would actually want to spend their time—even a few hours of it each day—on tedious control and accounting operations does not enter into the discussion here. In this section of Lenin's analysis, the need to organize state activity at all is merely a last necessary concession to time, and the state immediately begins to "wither away" along with time constraints themselves.

By contrast, the next section of *The State and Revolution*, entitled "The Economic Basis of the Withering Away of the State," insists on the proletarian dictatorship's need to utilize "bourgeois" state forms, and rational time discipline, for some time to come even under socialism. Directly contradicting his earlier assertions that the state after the revolution will be only a proletarian "semi-state" that would be in fact "no longer a state in the proper sense of the term," now, focusing on the likely postrevolutionary realities of economic production, Lenin suddenly declares that "under communism there remains for a time not only bourgeois law, but the bourgeois state, without the bourgeoisie!"[52] Here the transformative nature of revolution in effecting a break with time is relativized, and the "protracted" nature of the process of the state's withering away is stressed:

> The expropriation of the capitalists will inevitably result in an enormous development of the productive forces of human society. But how rapidly this development will proceed, how soon it will reach the point of breaking away from the division of labor, of doing away with the antithesis between mental and physical labor, of transforming labor into "life's prime want"—we do not and *cannot* know.
>
> That is why we are entitled to speak only of the inevitable withering away of the state, emphasizing the protracted nature of this process and its dependence on the rapidity of development of the *higher phase* of communism, and leaving the question of the time required for, or the concrete forms of, the withering away quite open, because there is *no* material for answering these questions.[53]

This means that the final overcoming of temporal constraints must be postponed until the "higher phase of communism" is reached:

> The state will be able to wither away completely when society adopts the rule: "From each according to his ability, to each according to his needs," i.e., when people have become so accustomed to observing the fundamental rules of social intercourse and when their labor has become so productive that they will voluntarily work *according to their ability*. "The narrow horizon of bourgeois law," which compels one to calculate with the heartlessness of a Shylock whether one has not worked half an hour more than somebody else, whether one is not getting less pay than somebody else—this narrow horizon will then be left behind.[54]

The ultimate ideal here remains the organization of production without any calculation of individual work time—but the implicit message is that in the interim period the need to keep track of labor and leisure time with all the "heartlessness of a Shylock" continues. Full communism and the overcoming of scarcity "presupposes not the present productivity of labor and *not the present* ordinary run of people"; meanwhile, "until the higher stage of communism arrives, the socialists demand the strictest control by society *and the state* over the measure of labor and the measure of consumption."[55] Despite Lenin's disclaimer that this is "by no means our ideal," his vision that under socialism "the whole of society will have become a single office and a single factory" with a single "factory discipline" contains little of the charismatic promise of a society beyond time constraints embodied in the Marxist vision of communism—and reiterated by Lenin in the first half of *The State and Revolution* itself![56]

Here we are confronted not with any synthesis between charismatic and rational understandings of time in Lenin's work but with a stark contradiction between them. The state after the revolution is held to be *both* "proletarian" and "bourgeois," based both on an absence of time constraints, allowing the participation of the masses in "increasingly simple" state control over social production, and on the enforcement of strict "factory discipline" to ensure that work time is used productively. In this light, it is not surprising that Lenin never returned to finish this confused and contradictory manuscript after the interruption of the October Revolution, explaining that "it is more pleasant and useful to go through the 'experience of the revolution' than to write about it."[57] In the end, all Lenin offered as a response to the dilemma of organizing state and economic institutions after the revolution was a restatement of Marx's split of communism into "lower" and "higher" stages—the one "bourgeois" and essentially subject to rational time, the other "proletarian" and essentially time transcendent—without showing any theoretical or empirical connection between the two.

After the Bolshevik revolution had been successfully completed, however, and questions of state and economic organization became unavoidable, Lenin's failure to arrive at a workable synthesis of charisma and rationality in dealing with issues of labor time had important consequences. As a substitute for a workable strategy for exercising state power in the postrevolutionary era, Lenin ultimately fell back on a strategy of merely safeguarding the one institution for which he had provided a coherent theoretical foundation—the party. The party, almost completely absent from the pages of *The State and Revolution*, was ironically to become the linchpin of the Soviet state in practice. This can be seen as a direct result of Lenin's inability to find an institutional formula for a Marxist "charismatic-rational" state that would parallel his formula for a charismatic-rational party of "professional revolutionaries." Such a parallel formula would elude the theoreticians of the Bolshevik Party until the victory of Stalinist economics in 1928.

Lenin's Postrevolutionary Writings and Political Activity

A full examination of the social factors leading to the political triumph of the Bolshevik Party in November 1917 in Russia would take us far beyond the bounds of this work. The ineptitude of the Provisional Government, the continuing devastation brought about by World War I, the mass peasant uprisings that accompanied the collapse of the tsarist autocracy, and the increasing radicalism of the working class were all significant factors in the Bolsheviks' success. The emphasis placed here on Lenin's ideological outlook should not be taken to imply that he single-handedly created the social conditions that allowed his party to come to power. Indeed, the truth is more nearly the opposite, as Lenin's political marginality prior to World War I demonstrates: Leninism could only have triumphed in an extremely turbulent social environment in which less radical political alternatives were quickly undermined.[58]

However, Lenin's unwavering conviction that the breakdown of the Russian autocracy heralded the beginning of the worldwide proletarian revolution, and that only the "party of professional revolutionaries" could successfully lead it, nonetheless did play a crucial role in determining the ultimate outcome of the Russian Revolution of 1917—and in shaping the formal political order institutionalized under Communist Party auspices in subsequent years. Lenin's decision to capitalize on the social radicalism engendered by state collapse, war, and economic crisis to launch the Bolshevik takeover has often been seen as the surest proof of Lenin's rejection of Marxism. Yet this decision was, as we have seen, based upon fifteen years of essentially Marxist theoretical argumentation and analysis. Moreover, the Marxist arguments against Lenin's decision raised at the time look surprisingly weak in retrospect.

The most common criticism of Lenin's seizure of power came from the Mensheviks and other "right" or reformist socialists: that scientifically, the productive forces of Russia had not "matured" enough for a socialist victory. Yet the argument that Marxists had to wait for "history" before beginning the proletarian revolution was the same excuse for inaction that had led the major parties of the Second International just three years earlier to acquiesce in the launching of the most devastating war in the history of humanity. By contrast, Marx and Engels had themselves argued in the *Communist Manifesto* that under conditions of "advanced civilization"—now globally in place, Lenin thought, as a result of the spread of imperialism—the jump from bourgeois revolution to proletarian revolution could be made "immediately." But surely eight months of capitalism was not sufficient for such a revolutionary leap?

Lenin's response, again, was consistent with Marx's: in times of revolutionary advance, one day might equal twenty years; only those hopelessly constrained by bourgeois notions of abstract time measurement could fail to perceive the qualitative change in how time was unfolding in the period after February 1917. Would it really have been more "Marxist" to reject Marx's own explicit position in the *Manifesto* and simply help to establish another century of capitalist rule in Russia?

Leftists, too, criticized certain aspects of the Bolshevik seizure of power. Luxemburg, while she welcomed Lenin's policy of pushing immediately for socialism in the revolutionary storm unleashed by the collapse of tsarism, remained troubled by what she perceived as Lenin's willingness to countenance hierarchical subordination of "spontaneous" worker socialism to a "bureaucratic" party elite. Once again, Lenin had only to point to the lessons of history to counter the argument: the Paris Commune had also been based upon the spontaneous organization of radical workers—and its lack of ability to defend itself against the capitalist state had doomed it to failure. Leftism, as Lenin continually insisted, was a form of childish romanticism that obscured the scientific analysis of real conditions and evaded the discipline necessary for revolutionary success. By 1919, with the failure of the Sparticist revolt in Germany and the executions of Luxemburg and Liebknecht, his argument appeared well founded.

Finally, the Bolshevik takeover was bitterly opposed by the symbol of centrist orthodoxy himself, the "renegade" Kautsky, who fumed at Lenin's violation of the "laws" of social development enshrined by Marx and Engels. "It is impossible in one bound to leap from Absolutism into a Socialist society," he insisted, thus reinforcing the Menshevik view.[59] Yet in addressing the question of what the truly "Marxist" response to the social upheaval in Russia might have been, Kautsky demonstrated a peculiarly orthodox ambivalence about the effects of revolutionary action:

> Consistent Marxism was thrown into a very difficult position when the Revolution set in motion the really great mass of the Russian people, who were conscious only of their needs and desires, and who did not care at all whether what they desired was, under the then [prevailing] circumstances, possible and socially advantageous. In the case of the Bolsheviks, Marxism had no power on the situation. The mass psychology overruled them, and they allowed themselves to be carried away by it. Doubtless in consequence of this they have become the rulers of Russia. It is quite another question what will and must be the end of it all.[60]

Thus Kautsky confessed that an actual social revolution posed grave "difficulties" for "consistent Marxism"—especially when those who most directly represented the revolutionary mood of the masses actually seized power! Apparently, Marxist orthodoxy for Kautsky now required a total detachment from the actual "needs and desires" of the workers, in order to mitigate the effects of "mass psychology." Predictably, Lenin rejoinders to this "centrist" criticism of Bolshevism were the most sarcastic and merciless of all.

In sum, the common view that Lenin "betrayed" Marxism in launching the Bolshevik revolution is not borne out by a careful analysis of the arguments against the Bolshevik coup made by various Marxists at the time. In historical context, Lenin's claim to have been a faithful Marxist in 1917 appears to be at least as good as those of his competitors. From a non-Marxist point of view, of course, it is very easy to criticize Lenin's establishment of what would eventually become one of the most tyrannical states in human history. But within the tradition of theorizing established by Marx and Engels, Lenin's positions and policies after the breakdown of tsarism make perfect sense.

After the revolution, Lenin continued to display the same combination of unwavering theoretical certainty about political issues, and confusion about economic and cultural issues, that had been characteristic of his work before 1917. On the one hand, concerning the question of preserving party power, Lenin showed a single-minded determination. No value, in Lenin's mind, could be higher than the preservation of the dictatorship of the workers' vanguard. Ideals of political pluralism were sacrificed to this principle with the dispersal of the predominantly non-Bolshevik Constituent Assembly, dreams of immediate global victory were sacrificed to it at Brest-Litovsk, and experiments with independent democratic decision making by the soviets and workers' collectives were gradually curtailed over the course of the Civil War. Finally, by 1921, remaining standards of inner-party democracy were sacrificed to preserve party unity in the face of the defection of formerly loyal soldiers and sailors during the Kronstadt revolt and the emergence of the Workers' Opposition within the party itself. As the hope of world revolution further receded, Lenin became even more convinced of the necessity of expelling any potentially contaminating, nonsocialist elements from the party. Within a year of Lenin's Ban on Factions at the Tenth Party Congress, the formal political order of the Bolshevik state began to look very much like the proposed structure of the Leninist party outlined in *What Is to Be Done*.[61]

As far as the organization of the socioeconomic activity of the masses was concerned, however, Lenin displayed not consistency but vacillation, arguing at different times both for increased centralized labor discipline and for the encouragement of spontaneous worker initiative in the raising of labor productivity. This vacillation—again, a reflection of the dichotomy between the rational and charismatic conceptions of time and state organization contained in *The State and Revolution*—can be illustrated by comparing Lenin's laudatory writings on two essentially contradictory approaches to labor organization: Taylorism and "communist *subbotniki*."

Lenin's interest in applying Taylorism—the principles of time-and-motion analysis as developed by the archcapitalist Frederick Taylor—to socialist conditions would be wholly inexplicable without an understanding of the rational component of the time orientation inherent both in Marxism and in Lenin's reinterpretation of Marxist theory.[62] The subordination of all worker activity, right down to the discrete motions of different parts of the body during factory labor to strict clock discipline, would seem to be the direct antithesis of the communist ideal—as indeed it appeared to many workers and intellectuals in the early Soviet period. Indeed, Lenin's first written reference to Taylorism, in March 1913, was essentially critical in tone.[63] Even after March 1914, when Lenin came out in favor of the application of Taylorist techniques under socialism, his emphasis on centralized labor discipline was muted by his simultaneous stress on worker self-government as laid out in the first part of *The State and Revolution*.

However, by the spring of 1918, after the conclusion of the Brest-Litovsk treaty, Lenin began to see the enforcement of work discipline as the crucial factor in the maintenance of Soviet power until the next revolutionary assault could begin. As the long-term nature of the transition to full communism after the revolution became clear, Lenin explicitly departed from his ideal of the commune-state expressed in the first part of *The State and Revolution* and insisted instead on the strict "accounting and control" over work emphasized in the second part: "What we are discussing is the shifting of *the center of gravity* of our economic and political work. Up to now measures for the direct expropriation of the expropriators were *in the forefront*. Now the organization of accounting and control in those enterprises in which the capitalists have already been expropriated, and in all other enterprises, advances *to the forefront*."[64] But Lenin recognized that the ultimate goal of stricter control over production—the raising of the productivity of labor time to capitalist levels and beyond—was an enormous task: "In this respect the situation is particularly bad and even hopeless if we are to believe those who have allowed themselves to be intimidated by the bourgeoisie. . . . The working out of new principles of labor discipline by the people is a very protracted process."[65]

It is in this context—combined with a frank admission of the failure of the revolution to transcend Russia's "backwardness" and hence its subjection to historical time—that Lenin introduces the argument for a "socialist" Taylorism:

> The Russian is a bad worker compared with people in advanced countries. It could not be otherwise under the tsarist regime and in view of the persistence of the hangover from serfdom. The task that the Soviet government must set the people in all its scope is—learn to work. The Taylor system, the last word of capitalism in this respect, like all capitalist progress, is a combination of the refined brutality or bourgeois exploitation and a number of the greatest scientific achievements in the field of analyzing mechanical motions during work, the elimination of superfluous and awkward motions, the elaborations of correct methods of work, the introduction of the best system of accounting and control, etc. The Soviet Republic must at all costs adopt all that is valuable in the achievements of science and technology in this field. The possibility of building socialism depends exactly upon our success in combining the Soviet power and the Soviet organization of administration with the up-to-date achievements of capitalism. We must organize in Russia the study and teaching of the Taylor system and systematically try it out and adapt it to our own ends. At the same time, in working to raise the productivity of labor, we must take into account the specific features of the transition period from capitalism to socialism, which, on the one hand, require that the foundations be laid of the socialist organization of competition, and, on the other hand, require the use of compulsion, so that the slogan of the dictatorship of the proletariat shall not be desecrated by the practice of a lily-livered proletarian government.[66]

In this passage we see all the elements of the rational, empirical Lenin in crystallized form—a clear realization of the low level of Russian labor discipline, the concomitant stress on the need for a systematic imposition of Western techniques of economic management, and the insistence that coercion would be necessary to bring this about.

However, during the summer of 1919, as the tide of the Civil War began to turn decisively in favor of the Red Army, Lenin began arguing for the increased utilization of a quite different form of labor organization: the so-called communist *subbotniki*.[67] *Subbota* means Saturday in Russian; a *subbotnik* was the voluntary donation by a brigade of workers of their Saturday day off for work in "communist construction." Whatever the actual motivations of those who originally organized and participated in these *subbotniki*, Lenin saw them as representing a truly revolutionary alternative to centralized forms of labor discipline—and a sign that the revolution itself was at last bringing about a new attitude toward labor. Here, the initiative to labor was being taken spontaneously by revolutionary workers who denied in practice the constraints on human activity in time implied by the need for periodic days of rest. In addition, such labor seemed to be extraordinarily productive. As Lenin describes one case, "Although the work was poorly prepared and organized the *productivity of labor was* nevertheless *from two to three times higher than usual*."[68] The possibility of extending this productivity—based not on compulsion but enthusiasm—to regular workdays seemed a promising one. As Lenin quoted a *Pravda* correspondent: "The comrades explain this [productivity] by the fact that ordinarily their work is boring and tiresome, whereas here they worked with a will and with enthusiasm. Now, however, they will be ashamed to turn out less in regular working hours than they did at the communist subbotnik."[69] While admitting that the party leadership could not "vouch that precisely the 'communist subbotniks' [would] play a particularly important role" in the further development of socialism, Lenin still recognized in this form of revolutionary labor organization the "tender shoots" of future communism: "Communism is the higher productivity of labor—compared with that existing under capitalism—of voluntary, class-conscious and united workers employing advanced techniques. Communist subbotniks are extraordinarily valuable as the *actual* beginning of *communism*; and this is a very rare thing, because we are in a stage when 'only the *first steps* in the transition from capitalism to communism are being taken' (as our Party Programme quite rightly says)."[70] The Soviet government, Lenin stressed, would do all it could to "nurture" these spontaneous tendencies toward truly communist methods of production—but how this could be reconciled with an emphasis on a socialist Taylorism enforced by the party was left out of the discussion.

Lenin never again displayed such unqualified enthusiasm for revolutionary work in communist *subbotniki*.[71] However, his analysis of their potential as incipient models of communism demonstrates that Lenin had not—at least in 1919—forgotten the charismatic promise of the time-transcendent commune-state outlined in the first part of *The State and Revolution*. Nonetheless, as the Civil War drew to a close in 1921 and persistent shortages of food and basic consumer goods in the cities along with a mass exodus of workers back to the countryside once again demonstrated the extent of the backwardness of the Soviet economy, Lenin logically fell back once again on a more pronounced emphasis on the need for centralized labor discipline as the proper socialist policy. With the proclamation of the New Economic Policy (NEP), with its emphasis on gradualism and slow improvements in the administration of economic work, Lenin's rational analysis of the organization of economic time overwhelmed any residual hopes for immediate charismatic transcendence in socialist economic production.

Lenin's reluctance to abandon all hope of further revolutionary advance can be seen in the incoherence of Lenin's attempts to provide a doctrinal justification of the NEP. Originally, Lenin argued that the implementation of the NEP represented in essence a revolutionary retreat. Then, in March 1922 Lenin demanded the "halting [of] the retreat" in preparation for a new period of advance. However, this period of advance was not to be based on new acts of revolutionary enthusiasm; the explicit task, as Lenin saw it, was to begin slowly and patiently to build up both administrative experience and labor discipline in Soviet institutions: "The retreat has come to an end; it is now a matter of regrouping our forces. These are the instructions that the Congress must pass so as to put an end to fuss and bustle. Calm down, do not philosophize; if you do, it will be counted as a black mark against you. Show by your practical efforts that you can work no less efficiently than the capitalists. . . . Even if the pace is a hundred times slower, it will be a million times more certain and more sure."[72] But then once again, in November 1922, he announced that a new "retreat" for "a new regrouping" had begun, while confessing that the party "did not know yet how to regroup."[73]

Despite Lenin's obvious confusion on how to characterize the NEP in terms of the advance to socialism, the struggle against "fuss and bustle" in economic and administrative work became Lenin's watchword. On the administrative level, Lenin pushed—as he had since 1917—for the establishment of stricter "control and accounting" over state activity through the agency of the Workers' and Peasants' Inspection (*Rabkrin*). And on the level of labor discipline, Lenin emphasized rational, centralized techniques of time management. Gone was any mention of the heroic exploits of laborers on communist *subbotniki*.

Lenin's "Testament"

Lenin's increasing incapacitation from 1922 onward precluded any further direct participation in the formulation of Soviet economic policy. However, between December 1922 and March 1923, he recovered sufficiently to write several important letters to the Central Committee—one containing his demand that Stalin be dismissed as general secretary of the party—and five last articles on questions of party composition and economic organization. These last five articles have sometimes been considered, by both Soviet and Western analysts, as Lenin's "testament" on these issues, codifying and extending his earlier analysis of the necessity of a long-term reliance on the NEP.

In fact, there is little that is new in these final essays as far as questions of economic organization are concerned. Lenin insists, even more strongly than before, on the protracted nature of the economic development of socialism and on the inadmissibility of acting in a hasty manner:

> We must show sound skepticism for too rapid progress, for boastfulness, etc. We must give thought to testing the steps forward we proclaim every hour, take every minute and then prove every second that they are flimsy, superficial and misunderstood. The most harmful thing here would be haste. The most harmful thing would be to rely on the assumption that we know at least something, or that we have any considerable number of elements necessary for the building of a really new state apparatus, one really worthy to be called socialist, Soviet, etc.
>
> No, we are ridiculously deficient of such an apparatus, and even of the elements of it, and we must remember that we should not stint time on building it, and that it will take many, many years.[74]

The deficiencies of the state apparatus are compounded, in Lenin's view, by the deficiencies in the cultural level of Russian workers: "The workers who are absorbed in the struggle for socialism . . . are not sufficiently educated. They would like to build a better apparatus for us, but they do not know how. They cannot build one. They have not yet developed the culture required for this; and it is culture that is required. Nothing will be achieved in this by doing things in a rush, by assault, by vim and vigor, or in general, by any of the best human qualities."[75] The practical solution to this problem given by Lenin here is, once again, a reliance on *Rabkrin*—purged of opportunist and corrupt elements—in the overseeing of state officials and such additional measures as the announcing of a "contest in the compilation of two or more textbooks on the organization of labor in general, and management in particular."[76]

The contrast with Lenin's vision of the commune-state in the first part of *The State and Revolution* and with his praise for communist *subbotniki* in 1919 could hardly be more stark. The only reference to charismatic time transcendence left in Lenin's analysis here is his sarcastic identification of "doing things in a rush" as one of the "best human qualities." Must we then accept the judgment that by 1921 Lenin had ceased to be a Marxist "Utopian" in any sense of the word and had settled on a pragmatic, essentially rational response to the problems of economic development in a backward socialist society?

Such a judgment, I would argue, is inadequate for an understanding of Lenin's theorizing and political activity in the postrevolutionary period. To begin with, one must bear in mind that Lenin's discussions of economic problems in general, and of economic time use in particular, were based on his conviction that the *political* problems of the Marxist revolutionary cause had been solved once and for all by the success of the Bolshevik Party in taking and holding on to state power. As he put it typically in "Better Fewer, but Better," "We . . . lack enough civilization to enable us to pass straight on to socialism, although we do have the political requisites for it."[77] Lenin's "pragmatism" in economic matters after 1921—that is, his subordination of the charismatic elements of his viewpoint to the rational ones—was possible only on the basis of his belief that the socialist character of the Soviet state was ensured as long as the "party of professional revolutionaries" he had created remained in charge.

On the issue of time use in economic production, an examination of Lenin's postrevolutionary writings, speeches, and policies taken as a whole reveals not a worked-out, consistent strategy but rather a continual vacillation between charismatic and rational approaches. The dichotomy between socialist time discipline and socialist time transcendence is presented in rather stark form in Lenin's only explicit theoretical work on the topic, *The State and Revolution*, and after that Lenin shifts his position from one emphasis to the other depending on his sense of the immediate prospects for revolutionary success. This cannot be seen as merely another example of Lenin's tactical flexibility—although he himself would almost certainly have justified it in these terms—because the two types of time use involved are antithetical and a reliance on one logically leads to the disappearance of the other. Promoting a societywide Taylorism would inevitably undermine the spontaneous enthusiasm for labor displayed in communist *subbotniki*; promoting communist *subbotniki* could not help but undermine strictly Taylorist approaches to labor discipline, which demanded a regularized, sober organization of time use. The paradox was thus similar to that faced by Bernstein and Luxemburg in the theoretical debate over political strategy: a "right" emphasis on respecting the need for bourgeois organization would lead to purely bourgeois success, while a "left" emphasis on revolutionary action would lead to a more socialist form of failure.

Lenin's last writings, in this respect, demonstrate more than anything else his desperate desire to find a solution to this paradox and his fear that failure to do so might lead to the spread of a purely "bureaucratic" and "Russian chauvinist" mentality such as that beginning to be displayed, in Lenin's view, by Joseph Stalin. Indeed, although the paradox of time discipline versus time transcendence remained unresolved in his economic policies, Lenin showed a keen insight into the dangers facing the revolutionary cause owing to the lack of a socioeconomic and cultural base for socialism in the Soviet state:

> Our opponents told us repeatedly that we were rash in undertaking to implant socialism in an insufficiently cultured country. But they were misled by our having started from the opposite end to that prescribed by theory (the theory of pedants of all kinds), because in our country the political and social revolution preceded the cultural revolution, the very cultural revolution which nevertheless now confronts us.
>
> This cultural revolution would now suffice to make our country a completely socialist country; but it presents immense difficulties of a purely cultural (for we are illiterate) and material character (for to be cultured we must achieve a certain development of the material means of production, must have a certain material base).[78]

This remarkable passage demonstrates Lenin's awareness of precisely what he had accomplished—a genuine political revolution, backed with social support—and what still remained to be accomplished before the achievement of socialism: first a socioeconomic revolution that would create the material basis for a new socialist attitude toward work, and then on that basis a cultural revolution that would lead to the internalization of this attitude by the masses. Ironically, it would be precisely Lenin's "bureaucratic" enemy, Stalin, who would successfully undertake the first of these tasks. The final attempt to complete the second, cultural revolution necessary for the establishment of a truly socialist society would be launched more than sixty years later, with the coming to power of Mikhail Gorbachev.

The Fragmentation of Political Leninism:
The Post-Lenin Succession and the End of the NEP

Lenin's incapacitation in 1923, followed by his death in January 1924, opened up the question of who would succeed the founder of the new Soviet regime. The ensuing political battle bore all the marks of spiritual and organizational struggle that are characteristic of all succession processes within decisively charismatic social orders. Indeed, the history of the struggle for power in the 1920s parallels the similar struggle between right, left, and center tendencies in the Second International after the deaths of the theoretical founders of the Marxist movement, Marx and Engels. Like the period of the prewar International, the Soviet 1920s saw the emergence of a more fully charismatic interpretation of the Leninist heritage, represented by Leon Trotsky, and of a more fully rational interpretation of that heritage, represented by Nikolai Bukharin. Once again there appeared a center faction resisting both left and right tendencies but with little positive program of its own beyond the preservation of orthodoxy—a tendency represented above all by Grigorii Zinoviev as the leader of the "triumvirate" of Zinoviev, Kamenev, and Stalin, which emerged as the most powerful faction in the Politburo during Lenin's final illness.

As during the Second International, none of these competing tendencies proved successful when confronted with the task of preserving the Marxist synthesis of charisma and rationality in time orientation—a synthesis that itself depended, as we have seen, upon the extension of "charismatic impersonalism" into new institutional realms. As with international Marxism in 1914–17, the last years of the 1920s in the USSR were marked by a sense among party members of decline and stagnation in the revolutionary cause—a decline reversed only with the emergence of a new innovator in the Marxist-Leninist tradition, Joseph Stalin, whose policies ushered in a new cycle in the development of the charismatic-rational conception of time. In all this, the battle of Trotsky, Bukharin, and Zinoviev, ending with Stalin's victory, parallels closely the earlier struggle of Luxemburg, Kautsky, and Bernstein before Lenin's revolutionary triumph in Russia.

However, in exploring these parallels we must also bear in mind three significant discontinuities between the periods of decline in the Second International and in the Soviet 1920s. First, the decline of "political Leninism" in the 1920s differs markedly from the analogous period in the previous "theoretical" cycle of development in that the success of Lenin's party of "professional revolutionaries" in taking and maintaining power in Russia provided a common political foundation for left, right, and center that was lacking in the Second International. In this sense, Trotsky, Bukharin, and Zinoviev had less in common with their analytic counterparts Luxemburg, Bernstein, and Kautsky and more in common with each other, simply because all three agreed on the fundamental necessity of preserving the political power of the Leninist party. This perceived necessity uniting the party elite worked to prevent what might otherwise have seemed a natural alliance between Trotsky's Leninist "left" and non-Leninist left-wing groups such as the Workers' Opposition, just as it prevented Bukharin from actively working to mobilize his potential nonparty constituency—peasants and the bourgeois technical intelligentsia—*against* the power of the party apparatus.

Secondly, the success of the Leninist party in maintaining state power created a new political environment in which the consequences of holding a defeated position were much more severe and irreversible than in the prerevolutionary period. Thus, whereas Bernstein, Luxemburg, and Kautsky all continued to exercise an important influence within the Marxist movement after experiencing theoretical defeat, Trotsky, Bukharin, and Zinoviev were denied any significant political role within the USSR after their expulsion from the leadership. Taken together, the Bolshevik elite's shared belief in the necessity of party power and party discipline, combined with the new institutional power of the party apparatus to suppress internal opposition to the party line, created a political milieu vastly different from that characteristic of the relatively fluid prewar Marxist movement—a change with implications that were recognized clearly by only one of the Bolshevik leaders, Stalin.

The third discontinuity between the period of struggle from 1923 to 1928 and the analogous period in the Second International was also a product of Lenin's political success: the task confronting the theorists of the 1920s had now shifted from that of finding a workable *political* institutionalization of charismatic-rational time to that of arriving at a synthesis of charismatic and rational time in *socioeconomic* affairs. Put simply, the question was: How could the "party of a new type" build a distinctively socialist "economy of a new type" in the USSR? This was, indeed, the problem that Lenin confronted in *The State and Revolution* but that remained fundamentally unresolved in his writings and in his policies until his death. Now, the end of the Civil War and the lifting of the immediate threat to Soviet Russia's survival meant that the question of economic construction necessarily moved to the fore.

Given these three major discontinuities between the period of the

1920s and the prewar International, it is all the more striking how closely the various responses of the major Bolshevik leaders to the new economic challenge parallel the trends of thought during the analogous period of revolutionary dissolution in the theoretical cycle after the deaths of Marx and Engels. Although left, right, and center tendencies within the Leninist elite did not emerge in the same chronological order as in the period of the Second International, all three tendencies *did* nonetheless emerge—each bearing in many ways a striking similarity to its prerevolutionary theoretical predecessor. It would thus not be too far wrong to think of Trotskyism as a sort of Luxemburgism, of Bukharinism as a sort of Bernsteinism, and of Zinovievism as a sort of Kautskyism—except that in each case these earlier tendencies were modified to include the paramount role of the Leninist party and to address more directly the question of "socialist construction" in the Soviet Union. In the section that follows, I examine the programs of Trotsky, Zinoviev, and Bernstein in historical sequence in order to make this argument in more detail.[79]

Trotsky's Charismatic Leninism

The first distinct tendency emerging with the fragmentation of political Leninism, even before Lenin's death, was the left tendency represented most consistently by Leon Trotsky. Trotsky's relationship to the Bolshevik Party (as well as to the Menshevik Party, to which he formally belonged until 1917), like Rosa Luxemburg's relationship to the German SPD, was marked by tension and often antagonism—an attitude to formal organization quite consistent with Luxemburg's and Trotsky's advocacy of a more fully charismatic interpretation of the Marxist revolutionary cause. Much of Trotsky's early theorizing, in fact, paralleled Luxemburg's in its stress on spontaneous revolutionary action as a purifying and time-transcending force. As Deutscher remarks, despite a certain lack of warmth in their personal relationship,

> of all the personalities of European socialism, nobody was in origin, temperament, and political and literary gifts more akin to Trotsky than Rosa Luxemburg. . . . Like Trotsky, she rejected the general Menshevik conception of the revolution, but viewed with suspicion the work of the Bolsheviks. Like Trotsky, she wanted to see the Russian movement "Europeanized," while she herself tried to breathe into the German party something of the Russian revolutionary idealism. They sometimes met at Kautsky's home, but they remained

aloof from each other, perhaps because of their extraordinary affinity. Agreeing so closely, they may have had little to say to each other.[80]

The affinity between Luxemburg and Trotsky extended as well to their common orientation toward revolutionary action over passive contemplation, as evidenced by Trotsky's central role in both the Russian Revolution of 1905 and that of 1917 and Luxemburg's unsuccessful attempt to foment revolution in Germany in 1919.

But although Trotsky, like Luxemburg, mistrusted Lenin's advocacy of organization over spontaneity, he was somewhat more willing than she to contemplate proletarian revolution as the work of a distinct minority of class-conscious workers, rather than view it solely as a mass phenomenon. This was due to his broader view of the nature of revolutionary transformation summed up in the famous phrase "permanent revolution." Trotsky's argument for permanent revolution, contained in his 1906 work *Balances and Prospects: Moving Forces of the Revolution*, was based on two assumptions: first, that after the experience of the Russian Revolution of 1905, history had advanced to the point where a new bourgeois revolution in Russia would inevitably be transformed into a socialist revolution by the mobilized urban workers, and second, that a socialist revolution in an Eastern country like Russia could not help but provide the spark for revolution in western Europe and indeed worldwide:

> The Russian proletariat . . . will meet with organized hostility on the part of world reaction and with readiness on the part of the world proletariat to lend the revolution organized assistance. Left to itself, the working class of Russia will inevitably be crushed by the counter-revolution at the moment when the peasantry turns its back upon the proletariat. Nothing will be left to the workers but to link the fate of their own political rule, and consequently the fate of the whole Russian revolution, with that of the socialist revolution in Europe. The Russian proletariat will throw into the scales of the class struggle of the entire capitalist world that colossal state-political power, which the temporary circumstances of the Russian bourgeois revolution will give it. With state power in its hands, with the counter-revolution behind its back, with the European reaction in front of it, it will address to its brothers all over the world the old appeal, which this time will be the call to the last onslaught: Proletarians of all lands, unite![81]

Believing that a revolution led even by a small minority of proletarians in Russia would be but the immediate prelude to the final victory of world socialism, Trotsky was less concerned than Luxemburg about the dangers of revolutionary elitism leading to a stifling of the creative initiative of the masses, and this made his eventual conversion to Leninism possible.

However, despite Trotsky's absolute acceptance of the Leninist ideal of the party after 1917, basic differences between Trotsky's and Lenin's interpretations of the revolutionary movement remained. Trotsky, at the core, never gave up the belief that proletarian revolution meant a continual forward march; he never fully internalized Lenin's conviction that to be a Marxist meant to accept a measure of "bourgeois" time discipline in pursuing revolutionary goals. The phrase "permanent revolution" itself conveys the charismatic essence of Trotsky's position when compared with Lenin's advocacy of "professional revolution"; "permanent revolution" implied a continuous process of time transcendence, whereas "professional revolution" implied a struggle against time governed by an understanding of the constraints still imposed on revolutionary action by impersonal temporal forces.

It would thus be incorrect to argue that Trotsky's views became wholly compatible with Lenin's after 1917, just as it is incorrect to argue (as Trotsky himself rather immodestly did) that Lenin simply adopted Trotsky's earlier ideas about permanent revolution in calling for revolutionary action in October. By siding with Lenin in the Bolshevik revolution, and by devoting his considerable talent as a political and military organizer to Lenin's cause, Trotsky became a crucial actor in the establishment of political charismatic rationalism and thus departed from the purer revolutionary romanticism of Luxemburg on political issues. But, as one can see in his arguments on how to understand the nature of socialist economics and of socialist culture, Trotsky in important ways remained true to his prerevolutionary left, charismatic interpretation of Marxism.[82]

The first evidence of Trotsky's differences with Lenin on issues of economic organization came in 1920, in the debate over Trotsky's proposal for the "militarization of labor." As head of the Red Army, Trotsky was in a perfect position to act on his conviction that true revolutionary discipline meant a continuous, orderly, collective assault. Indeed, as long as the war continued, there could be no perceived conflict between the need for strict discipline and a sense of common participation in the revolutionary struggle. Naturally, then, when the war had substantially been won, Trotsky sought to apply the methods of military organization to the task of socialist economic construction. Accordingly, he began to call for the formation of "labor armies" and the strict subordination of the trade unions to military discipline.[83]

Trotsky's advocacy of labor militarization should by no means be seen as contradicting his emphasis on the role of revolutionary enthusiasm in building socialism. Instead, Trotsky's economic program in 1920 can be understood as an attempt to outline a party-led form of Luxemburgian mass revolutionary action in economic life. Like Lenin, Trotsky advocated the application of the "progressive" elements of Taylorism to Soviet work; but unlike Lenin, Trotsky never combined this advocacy of Taylorism with arguments for a slower tempo of socialist construction, nor did he see any contradiction between stricter labor discipline and quick tempos of revolutionary advance. Taylorism and military discipline in labor were for Trotsky only tools with which the party could direct the continuous forward movement of the socialist cause on the home front: "Just as we once issued the order 'Proletarians, to horse!', so we now must raise the cry 'Proletarians, back to the factory bench! Proletarians, back to production!' "[84] Those who avoided labor discipline, Trotsky thought, would naturally be treated by their revolutionary comrades at work as the moral equivalents of wartime deserters, who had subverted the general enthusiasm of the proletarian labor army:

> Display untiring energy in your work, as if you were on the march or in battle. . . . Commanders and commissars are responsible for their detachments at work as in battle. . . . The political departments must cultivate the spirit of the worker in the soldier and preserve the soldier in the worker. . . . A deserter from labor is as contemptible and despicable as a deserter from the battlefield. Severe punishment to both! . . . Begin and complete your work, wherever possible, to the sound of socialist hymns and songs. Your work is not slave labor but high service to the socialist fatherland.[85]

Unfortunately for the feasibility of Trotsky's vision, by 1920 Soviet workers had had more than their share of wartime mobilization. After some early success in reorganizing the Central Transport Commission, Tsektran, Trotsky's methods came under severe attack. The trade unions bitterly criticized his plans for their complete subordination to central party directives, and Lenin dissociated himself from Trotsky's ideas on this subject. The resulting ill will borne by workers' representatives toward Trotsky created barriers between him and left-wing opposition groups such as the Workers' Opposition, and this effectively deprived Trotsky of important potential allies in his later battles with Zinoviev,

Bukharin, and Stalin. Trotsky's attempt to reconcile a strict, hierarchical direction of labor by the party leadership with a leftist emphasis on charismatic rather than rational attitude toward work was ultimately a failure, for this strategy depended upon Trotsky's fundamental prior assumption that the revolutionary victory of the Leninist party had begun a process of world revolution from which there could be no turning back and which would suffer no reversals. The waning of proletarian enthusiasm in Russia after the Civil War and the simultaneous waning of the revolutionary movement worldwide thus meant the loss of the coherence of Trotsky's position.

From the introduction of the NEP onward, Trotsky's economic proposals were marked by a certain conceptual ambivalence—if not outright incoherence. On the one hand, Trotsky continued as he had since the early days of the revolution to be a staunch advocate of central planning on a nationwide level—a policy that Lenin had opposed as premature for a backward country like Russia. Together with Preobrazhensky, Trotsky developed the argument for a "primitive socialist accumulation"—implicitly at the expense of the workers as well as the peasants—to provide the basis for a faster tempo of industrialization and a quicker forward march for socialism.[86] In all this, Trotsky remained as before an advocate of centralization and hierarchical discipline in economic matters. On the other hand, Trotsky began to attack what he saw as the increasing bureaucratization of the party apparatus, the only cure for which would be the revolutionary participation of the masses in socialist construction rather than the passive carrying out of orders from above. But his struggle against bureaucracy would seemingly have required Trotsky to argue *against* centralized decision making by the party rather than for it—a step he was never willing explicitly to take. Rather, he continued to place his faith in the idea that strict party control—under the right sort of leadership—would somehow inspire the socialist masses to continuous revolutionary activity.

The year 1923 represented the last point at which Trotsky could articulate his views from within the political elite. Yet his contradictory advocacy of the preservation of Leninist discipline from above and the mobilization of mass revolutionary enthusiasm and initiative from below placed him in a politically impossible position. Trotsky's calls for a less stifling institutional order within the Bolshevik Party contrasted uneasily with his reputation as the party's most stern disciplinarian, which opened him up to charges of hypocrisy by the party center, led in Lenin's absence by the triumvirs Zinoviev, Kamenev, and Stalin. To avoid the charge of being insufficiently loyal to the party, Trotsky felt compelled to dissociate himself from his closest supporters, such as the signers of the so-called Platform of the Forty-six—a document that echoed Trotsky's calls for economic planning and party democracy but in terms too directly opposed to the party's official line for Trotsky to endorse explicitly. The end result was that Trotsky became almost completely isolated politically.

Ultimately, Trotsky's stress on revolutionary action over bureaucratic politics reflected his personal distaste for the infighting and maneuvering necessary to succeed in the postrevolutionary political milieu. Despite Lenin's urgent request that Trotsky make public the dying leader's testament urging Stalin's removal as general secretary, Trotsky at the critical point in the power struggle did nothing, asking instead to be allowed to travel to Germany to help prepare the attempt at proletarian revolution there. As Deutscher puts it, "To contribute once more to the victory of a fighting revolution suited him better than to taste the maggoty fruit of a victorious one."[87] But this attitude toward institutionalized politics—charismatic to the core—proved fatal to Trotsky's opposition to the party center.

Trotsky thus remained, in crucial respects, a revolutionary romantic. Despite his occasional attempts to formulate workable alternatives to the center's economic and foreign policies, he remained wedded to a vision of a triumphant world socialist movement of which the Russian Revolution was but the prelude. The discipline he was interested in was discipline for the forward march of the movement, not discipline for its own sake, as itself something to be positively valued. Trotsky's ultimate standard of revolutionary success was a world in which "man will grow incomparably stronger, wiser, subtler; his body will become more harmonious; his movements more rhythmical; his voice more musical. The forms of his existence will acquire a dynamic theatrical quality. The average man will rise to the stature of Aristotle, Goethe, Marx. And above these heights new peaks will rise."[88] Compared with this standard—a world in which everyday life would become inherently "dynamic," unchained by the dead weight of time—present-day, mundane politics could have no appeal. The best that could be said of party activity under the NEP was that it might prepare the ground for a further stage of triumphant revolutionary advance: "We are soldiers on the march. We have a day of rest. Our present . . . cultural work is but an attempt to bring ourselves into some sort of order between two battles and two marches."[89] But such a conception of the NEP seriously underestimated the time required before the possibility of a new forward march and, more crucially, could provide no guidance for dealing with the "day of rest" itself in a revolutionary manner. As from the beginning of his political life, Trotsky could not bring himself to see restraint, as well as action, as a necessary component of revolutionary discipline.

As the Bolsheviks' "day of rest" stretched into years, therefore, Trotsky became less and less able to formulate a workable programmatic alternative to what he perceived as the stagnation of the revolutionary movement. At the same time, however, he continued to profess his loyalty to the party in almost overly emphatic terms. Trotsky's ambivalence about his role as leader of the main opposition to the party line is evident in his response to Zinoviev's call for a recantation of his views at the Thirteenth Party Congress in May 1924:

> Nothing could be simpler or easier, morally and politically, than to admit before one's own party that one had erred. . . . Comrades, none of us wishes to be or can be right against the party. In the last instance the party is always right, because it is *the only historic instrument which the working class possesses for the solution of its fundamental tasks.* . . . One can be right only with the party and through the party because history has not created any other way for the realization of one's rightness.[90]

This combination of programmatic weakness with political inhibition doomed Trotsky's activities against the center from 1923 on. At critical points in the power struggle, such as Stalin's breaking of Zinoviev's hold over the Leningrad party organization in 1925, Trotsky did nothing. When Trotsky did finally ally himself with Zinoviev and Kamenev in the United Opposition of 1926–27, he did little but repeat the left's earlier charges of bureaucratic deformation within the party and the need for a greater emphasis on industrialization, proposing no feasible domestic economic policy to counter Bukharin's. (Stalin's independent ability to formulate Bolshevik policy, even at this late stage, was almost universally underestimated.)

After the defeat of the United Opposition, Trotsky was arrested and ultimately exiled from the Soviet Union. Ironically, it was his final expulsion from day-to-day political life in the Leninist regime that rekindled his revolutionary enthusiasm, allowing him at last to be free to play the role of the revolutionary crusader—now against Stalinism as well as against capitalism—without being tied down to the uncomfortable demands of mundane power politics. The failure of Trotsky's charismatic Leninism ultimately led him, in his last years, to what was in essence a return to the Luxemburgian, purely theoretical charismatic interpretation of Marxism: preaching the purifying power of revolution without any ability successfully to make one.

Zinoviev and Leninist Orthodoxy

The role of Grigorii Zinoviev in the early years of the Soviet period has been quite neglected by scholars. Despite Zinoviev's role as Lenin's right-hand man in the years between 1910 and 1917, and despite the fact that during the critical period between Lenin's first stroke in 1923 and Stalin's alliance with Bukharin in 1925 Zinoviev was widely perceived to be the single most important man in the Bolshevik Party, there is not yet a single full-length biography of him. I can hardly hope to rectify this problem in a few pages; much of what I explore here is therefore somewhat speculative. Nonetheless, I will argue that what historians of the Soviet 1920s have seen only as evidence of Zinoviev's weakness of character—a certain dogmatism in defending Leninist orthodoxy, combined with vacillation on economic policy—are in fact traits consistent with a centrist, orthodox interpretation of political Leninism.[91] Zinoviev, in short, was attempting to play a role similar to that of Kautsky in the Second International, as defender of the faith and interpreter of the founder's ideas. That he did so rather unsuccessfully does not diminish the importance of the attempt or the necessity of making analytic sense of it, for in establishing, at least in principle, the concept of a "Leninist center" opposed both to left and right "deviations," Zinoviev created the formal political base of Stalinism—though Stalin would, in 1928–29, reinfuse that formal base with new revolutionary content.

Just as Kautsky had relied on his intimate relationship with Engels to validate his implicit claim to be the embodiment of Marxist orthodoxy during the period of the Second International, Zinoviev's close relationship with Lenin during the lowest ebb of the latter's fortunes in the prerevolutionary years—when practically no other leading Bolshevik stood by the future revolutionary leader—was the major basis of his claim to legitimate authority. Like Kautsky, Zinoviev worked to build a reputation as a theorist of the orthodox center, culminating in the publication of his book *Leninism* in 1925.[92] Finally, just as Kautsky was the recognized leading theorist of the international Marxist movement, Zinoviev's major party post was his position as leader of the Third International, founded in 1919—a position of enormous prestige at a time when world revolution was thought to be imminent.

During the period of Lenin's illness and shortly after his death, Zinoviev took advantage of this background in order to take de facto control of the party. This may well have been, as Carr has pointed out, a foolish decision, since it allowed Stalin to remain quietly in the background as Zinoviev battled Trotsky.[93] But such criticism is easy to make in retrospect. Given the opportunity to inherit Lenin's mantle, Zinoviev not surprisingly jumped at the chance; indeed, it would have required a remarkable degree of perspicacity to foresee the dangers involved.

And Zinoviev was a quite unremarkable politician, as well as an average thinker—much as Kautsky had been a solid, but hardly brilliant, interpreter of the work of Marx and Engels. As Carr points out, quoting a contemporary observer, "if 'what [Zinoviev] wrote was neither deep nor original' it was always serviceable."[94] Since he also apparently possessed excellent speaking abilities, Zinoviev was well suited to his role of a defender of political orthodoxy, even if he was quite ill suited to the task of bringing about the construction of the world's first socialist state. In short, he was an ordinary man who had the misfortune of being judged by the extraordinary standards appropriate to so many of his contemporaries.

Thus, Carr excoriates Zinoviev for being the initiator, during the struggle of the triumvirate against Trotsky and the Left Opposition, of such future Stalinist practices as branding all criticism of the party line as "objectively Menshevik criticism," calling upon defeated opponents such as Trotsky to criticize *themselves* publicly after their defeat, and helping Stalin build up a quasi-religious cult around the dead Lenin.[95] In fact, with the exception of the practice of public self-criticism, none of this was truly new in the history of the Marxist movement. Before 1917, Lenin himself had certainly never shied away from using the most damning possible language to lambast "deviations" from Marxism, and the practice of using quotations from Lenin in a theological rather than analytic way merely replicated the tendency of Marxists during the period of the Second International to use quotations from Marx and Engels in a similar manner. The reality of Bolshevik state power and the increasingly coercive relationship of the party to its defeated opponents gave Zinoviev's applications of old methods of Marxist debate new, potentially totalitarian implications—but we must blame this not so much on Zinoviev's sinister character as on the unforeseen effects of his fidelity to an older tradition of Marxist discourse in changed political circumstances.

Zinoviev's political platform can, in fact, best be analyzed as a sort of

Leninist religious orthodoxy, comparable in its ethos to the Marxist theoretical orthodoxy of Kautsky. According to Zinoviev, the Bolshevik Party, and the Soviet state it presided over, was to be the symbol of revolutionary purity until the revolution itself broke out on a worldwide scale, the concrete manifestation of charismatic communism in the mundane world—just as the SPD had represented a concretization of the mass proletarian revolutionary movement for Kautsky. And just as Kautsky had claimed to provide an accurate interpretation of the will of the founders of scientific socialism, Marx and Engels, Zinoviev saw his stewardship over the party as sanctified by his fidelity to the founder of the Bolshevik regime, Lenin.[96] In short, in place of the disciplined revolutionary movement, Zinoviev sought to substitute a disciplined revolutionary church, relying on a quasi-theological, analytically traditional form of political authority that had plenty of precedent in Russian history and in the prewar international Marxist movement. The decision to embalm Lenin's body and place it on permanent display in Red Square, which Zinoviev supported enthusiastically, was perhaps the most concrete example of this substitution of religious symbolism for revolutionary charisma.[97]

Zinoviev's seeming lack of an economic strategy for building socialism, like his political outlook, can also be seen not merely as a reflection of his personality but as consistent with a centrist, orthodox interpretation of Marxism-Leninism. Zinoviev's economic policy, like the political policy of the German SPD during the period of Kautsky's theoretical dominance, could be described as one of "revolutionary waiting" for an extended period: "How many years can the proletarian power maintain itself in one country if the proletarian revolution fails to occur in the others? At the outset, we had no precise ideas on the subject. It's only now that we have begun to estimate precisely the 'time' factor."[98] But since, as Lenin had pointed out, there was "no law according to which countries must live according to the Bolshevik revolutionary calendar," the precise amount of time necessary before the final achievement of socialism could not be known in advance.[99] In the meantime, then, the party should simply preserve the "link between the peasantry and the proletariat" under the auspices of the NEP, while fighting off both left and right misinterpretations of this process:

> Every time Bolshevism must undertake a strategic retreat, it inevitably encounters people, groupings, tendencies, factions and currents who, not understanding the historic meaning of the strategic maneuver in question, announce that the Bolsheviks have turned to the

right, that they are presiding over the degeneration of Bolshevism, the twilight of communism, etc.

The left addresses these reproaches against Leninism with bitterness and sincere despair. The right rejoices at this so-called revolution. . . . Neither one nor the other understands Leninism. They do not understand that the party of the proletarian revolution can remain faithful to itself in spite of any sort of strategic retreat.[100]

But in practice, Zinoviev's desire to avoid "left" and "right" deviations in economic policy led him simply to vacillate between Bukharinist and Trotskyist policies, first proclaiming "Face to the Countryside!" as the party's slogan, then later backing the demands of heavy industry and the Leningrad workers against the policies of the Bukharin-Stalin duumvirate.

In this respect, though, Zinoviev differed little from Lenin, who was similarly at a loss as to how to reconcile the conflicting demands of revolutionary construction and the maintenance of power in a predominantly peasant country. It had been Lenin, after all, who had argued that the main guarantee of the socialist nature of the Soviet state was the simple maintenance of the political dictatorship of the party. One could argue that Zinoviev's passionate defense of party discipline and orthodoxy, combined with his fundamental dualism on economic questions, thus represented a more faithful continuation of Lenin's heritage than either Trotsky's romantic Leninism or Bukharin's evolutionary socialism, which is examined below.

In one crucial sense, however, Zinoviev did depart from the essence of Leninism, just as Kautsky had departed from the essence of Marxism. That essence was, as I have argued throughout this work, the synthesis of the revolutionary collapsing of time with rational time discipline that made Marx's and Lenin's vision of revolution so compelling and powerful—a synthesis that, as the experience of the Second International had shown, could not fail to degenerate in the absence of any opportunity for practical revolutionary action. This Zinoviev had little stomach for, as his opposition to and public betrayal of Lenin's plans for political takeover in 1917 had demonstrated.

Considering the devastation of the Soviet economy in 1924, Zinoviev's desire merely to maintain the purity of revolutionary discipline attained by the Bolshevik Party under Lenin, without risking everything through a fresh assault on Soviet society, was in a way quite understandable. If it had not been for the presence in the party of men superior to him in matters of economic theory, such as Bukharin, and of men superior to him in matters of theory *and* of practical politics, such as Stalin, Zinoviev might well have presided over the Bolshevik Party for much longer than the two highly significant years he did do so. Had this occurred, the party would no doubt have degenerated over time—but there was no inexorable force of history forcing the Bolsheviks on toward further revolutionary battles. Indeed, it is only when we begin to see Zinoviev as a possible third alternative to Stalin, in addition to Trotsky and Bukharin, that we may begin to understand the appeal of a later leader who presided over a quasi-theological, stagnant, and degenerate form of Leninism—Leonid Brezhnev.

Bukharin and the Right Opposition

The revisionist trend in Soviet historiography has played a crucial role in restoring Nikolai Bukharin to a central place in the struggles of the 1920s, correcting the tendency of earlier historians such as Carr and Deutscher to see Trotsky as the only serious alternative to Stalin. However, certain authors have tended to combine their welcome reemphasis of Bukharin's role with an additional, more dubious argument—that his policies were the most consistent with Lenin's of any of the contenders for power in that period. Reversing the Carr-Deutscher argument entirely, for example, Stephen Cohen dismisses Trotsky as a hopeless romantic and holds out Bukharin as originator of a realistic socialist alternative:

> For all of his achievements as a leader and oppositionist, Trotsky never managed to develop in the 1920s a distinct, coherent set of policies for industrialization and socialism in Soviet Russia. Nor did his diffuse ideas and flashes of insight exercise mass appeal inside or outside the party. Bukharin, on the other hand, despite his inadequacies as a politician, became the leading spokesman for specific ideas and policies—the principles and practices of NEP—that were both a premonition of and alternative to Stalinism. They had considerable appeal in the party and in the country, before and after Bukharin's defeat. And they did not prove to be an "inherent impossibility"; they were forcibly abolished, along with NEP.[101]

I will argue, however, that despite its practical advantages, Bukharin's program was as unsuccessful as Trotsky's in reconciling the charisma-rationality dualism at the heart of Marx's and Lenin's work. Whereas Trotsky remained unable to compromise his conception of revolution as a purely charismatic force by allowing for periods of revolutionary patience, Bukharin counseled patience at the expense of the charismatic essence of the Bolshevik cause. Bukharin, despite his wholehearted endorsement of Lenin's conception of the party, thus represents the political equivalent of Bernstein's theoretical revisionism in the political cycle of Leninist development.

Bukharin's early career as a revolutionary would hardly seem to portend his eventual emergence as the primary articulator and defender of a form of evolutionary socialism in the USSR. Like Bernstein, Bukharin began his career on the extreme left of revolutionary movement.[102] Just as Bernstein had formed a close bond with one of the founders of theoretical Marxism, Engels, Bukharin forged a close personal relationship with the founder of Bolshevism, Lenin. Although openly disagreeing with Lenin on certain fundamental issues where Bukharin stood with the left, such as the national question, he nonetheless looked upon the Bolshevik leader with intense admiration and affection. Finally, like Bernstein, Bukharin at a certain point switched from left to right and became the most consistent advocate of what we might term a rational-legal interpretation of Leninist economic development, opposed to all attempts at forcing the pace of history; in this respect his views parallel Bernstein's theoretical advocacy of evolutionary socialism within the confines of the bourgeois social system.[103]

Bukharin's role as advocate of the extreme left viewpoint in the party continued for a few years after the October Revolution—notably during Brest-Litovsk, when he was one of the primary advocates of revolutionary war against Germany, and during the period of war communism, when he stood with Trotsky for the formation of labor armies, statization of the entire economy, and a rapid transition to full socialism. But by the end of 1921, three factors led Bukharin to reevaluate his left position: the collapse of war communism and Lenin's introduction of the NEP, Bukharin's increasing dissatisfaction with the economic program of Trotsky and his supporters as it became more clearly defined, and certain elements of Bukharin's theoretical understanding of the nature of the transition to socialism that inclined him toward a more gradualist viewpoint in conditions of social stabilization.

It is the last of these factors that we must analyze in depth in order to understand why the first two affected Bukharin's thinking as they did. What was the difference between Bukharin's leftism and Trotsky's that led these erstwhile allies, looking at the same empirical situation, to adopt such diametrically opposed positions on economic policy in the

1920s? In fact, the similarity between the two types of leftism had all along been more apparent than real. Trotsky, like Luxemburg, centered his analysis on a charismatic conception of revolutionary action as a heroic, continuous battle against time; his special difficulty in dealing with the problem of socialist construction in a postrevolutionary regime was his inability to reincorporate temporal constraints into his vision of the socialist order. Bukharin, by contrast, tended at first to see socialism as a stable, enduring Utopia that would be established after a relatively short period of revolutionary struggle. In a sense, then, Bukharin's difficulty in dealing with the problem of socialist construction was precisely the opposite of Trotsky's. While Trotsky could not find a way to justify rational time discipline under socialism, Bukharin had no way to reincorporate the charismatic collapsing of time into his conception of socialist development after the failure of the October Revolution and Civil War to bring about immediate and complete social transformation. Trotsky wanted "permanent revolution"; Bukharin wanted the revolution to come to a quick, successful conclusion. When this did not occur, Bukharin was forced to reformulate his conception of socialism in terms compatible with what now appeared to be a lengthy period of mundane, linear time stretching out indefinitely into the future.

Bukharin's switch from Utopianism to reformism was thus a logical one. This can be clearly seen by contrasting the two very different, yet related, works written by him in 1921: the first, *Economics of the Transition Period*, a paean to war communism and a prediction of the rapid emergence of a communist order; the second, *Historical Materialism*, an expression of Bukharin's new evolutionary thinking about socialism. At first glance, one might think the second book completely contradicted Bukharin's earlier point of view. And yet the two treatises had one important feature in common: namely, a stress on the concept of *equilibrium* as the normal state of affairs in all social systems—capitalist and socialist alike.

Bukharin's notion of equilibrium was not derived from mainstream Marxism but rather from his study of bourgeois sociology—an interest of Bukharin's that Lenin characteristically subjected to merciless ridicule.[104] Indeed, Bukharin's conception of equilibrium, even in the *Economics of the Transition Period*, did have a decidedly "bourgeois" sociological flavor to it:

In a society with a social division of labor . . . there must be a certain *equilibrium* of the whole system. The necessary quantities of coal, iron, machines, cotton, linen, bread, sugar, boots, etc., etc., are being produced. Living human labor is expended in accordance with all of this in the necessary quantities in relation to production, utilizing the necessary means of production. There may be all sorts of deviations and fluctuations, the whole system may be enlarged, complicated, and developed; it is in constant motion and fluctuation, but, in general and in its entirety, it is in a state of equilibrium.

To find the law of this equilibrium is the basic problem of theoretical economics.[105]

Bukharin's application of the concept of "equilibrium" to Marx's conception of capitalism was questionable, since Marx would no doubt have objected to any notion of capitalism "in general and in its entirety" that did not refer to its necessary revolutionary destruction by the proletariat. But the implication here that *socialism* as well had to reach a "necessary" functioning equilibrium, with "necessary" expenditures of labor, utilizing the "necessary means of production," displayed a fundamental misunderstanding of Marx's charismatic vision. For Marx—at least in his "revolutionary" mode—socialism was different from capitalism precisely because it pointed the way to a world *beyond the realm of necessity*. Whereas the realization of Marx's vision of socialism depended on an overcoming of the human subordination to time in order to abolish scarcity, Bukharin's notion of socialist equilibrium left socialist society in the same historical boat as capitalist society, trying merely to find a mechanism of "equilibrium" that would manage continuing scarcity more or less effectively for a more or less lengthy period.

It is true that in the *Economics of the Transition Period* Bukharin argued that the concept of equilibrium does not apply during the period of the revolutionary smashing of the old order. Yet a careful examination of Bukharin's reasoning on this score betrays the fundamental time-boundedness of the whole "equilibrium" approach:

> *The postulate of equilibrium* is invalid [in the revolutionary period]. . . . There is neither proportionality between production and consumption, nor between different branches of production . . . nor between human elements of the system. Therefore it is radically wrong to transfer to the transition period categories, concepts, and laws adequate to a state of equilibrium. One may object that insofar as society has not perished, there is equilibrium. Such reasoning, however, would be correct if the period of time we are examining was conceived of as being of great length. A society cannot live *long* outside equilibrium, it dies. But this social system for a certain time can be in an "abnormal" state, i.e., outside a state of equilibrium.[106]

Bukharin's conception of revolution as being an "abnormal" period of mundane time, rather than a vehicle for the transcendence of time, could hardly be illustrated more clearly. If revolution inherently destroys equilibrium, and if "a society cannot live *long* outside equilibrium" before it dies, then according to Bukharin a "revolutionary society" is by definition impossible. Yet the building of consistently revolutionary socioeconomic institutions was precisely what was demanded if the synthesis of charismatic and rational time orientations at the core of Marxist and Leninist thought was to be preserved in the Bolshevik regime. As Cohen points out, the above passage could be interpreted only in one of two ways: "Either the transition to socialism would be relatively brief; or Bukharin meant only the transition to a stabilized state from which socialism would evolve. It is reasonable to assume that in 1920 he believed the former. After 1921, however, he offered the second interpretation."[107]

However, neither of the two sides of Bukharin's thinking maintained a consistently charismatic-rational conception of time; both are based fundamentally on an acceptance of time as linear and inexorable. Bukharin's early work assumed that Utopia would emerge at a concrete point on the historical time line after the revolution. By the end of 1921, Bukharin had rejected his earlier hopes as "childish illusions" and began to see the achievement of Utopia as a process of linear evolution requiring decades. In neither case did Bukharin allow any permanent role for revolutionary communism as "the real movement which abolishes the present state of things"; in both periods he portrayed revolution as a delimited historical event rather than as a vehicle for the continuous collapsing of time. Having concluded that the October Revolution had failed to bring about the end of history, Bukharin had no choice, given his theoretical outlook, but to see socialist society as finally and absolutely subordinated to the rule of time.[108]

Accordingly, Bukharin in *Historical Materialism* began to argue that socialist society, like capitalist society, had to find mechanisms for ensuring a "new and durable envelope of production relations . . . capable of serving as an evolutionary form of the productive forces."[109] By 1922, Bukharin, like Bernstein before him, began explicitly to advocate a theory of "growing into socialism"—one he distinguished from Bernstein's revisionism by stressing that this process could take place only in a postrevolutionary society such as the USSR, never in a capitalist order:

We shall not be able to fulfill our task by single decrees, by single compulsory measures. . . . A prolonged organic process . . . a process of real growing into socialism will be required. But the difference between them and us is establishing when growing in begins. The revisionists, who do not want any kind of revolution, maintain that this process . . . occurs already in the bosom of capitalism. We maintain that it begins only together with the dictatorship of the proletariat. The proletariat must destroy the old bourgeois state, seize power, and with the help of his lever change economic relations. We have here a lengthy process of development, in the course of which socialist forms of production and exchange obtain an ever wider dissemination and, in that way, gradually displace all the remnants of capitalist society.[110]

This remained the essence of Bukharin's position throughout the 1920s: the continuation of the "dictatorship of the proletariat," which had been concretized as the rule of the Bolshevik Party, would guarantee that Soviet economic development would remain essentially "socialist" in nature; socialist development itself, however, would be a time-bound, "prolonged organic process."

Bukharin's theoretical position from 1921 on can therefore justifiably be termed a sort of "Leninist Bernsteinism." Like Bernstein's revisionism, Bukharinism had much to recommend it on the level of practical effectiveness. Having conceded the impossibility of mastering time during the prolonged transition to socialism, Bukharin was free to construct an economic platform free from many of the inconsistencies of Trotsky's or Zinoviev's policies. But, like Bernsteinism, Bukharinism's credentials as a truly Marxist-Leninist, revolutionary approach to building socialism were suspect, for as the concrete economics of Bukharin's evolutionary theories began to be implemented from 1925 to 1928, the implications of accepting the inexorability of time in the USSR became clear. As Bukharin insisted that socialism would emerge from the present-day Soviet system only "at a snail's pace," and as a quasi-capitalist time discipline began increasingly to be imposed on Soviet workers in the name of the "rationalization" of industrial activity in the USSR, the claim of Bukharin to be following in Lenin's footsteps began to sound increasingly hollow. What sort of Marxism or Leninism, after all, proclaimed revolution to be a thing of the past, while calling for a long-term continuation of an NEP that appeared to many Soviet workers as nothing more than a "New Exploitation of the Proletariat"?[111]

Cohen is quite correct to insist that Bukharin's policies "did not prove to be an 'inherent impossibility.'" If anything, the possibility of the party-state presiding over a pro-peasant policy of gradual industrialization was, from the perspective of many party members and workers, all too real. Unfortunately for Bukharin, such "realism" both departed from the charismatic essence of the time orientation of the Marxist-Leninist movement and went against the material interests of the crucial segment of urban workers who provided the backbone of social support for the new Soviet regime. By themselves, these factors might not have been enough to bring about Bukharin's downfall. But with Joseph Stalin's articulation in 1928–29 of a distinctive set of socioeconomic policies that seemed both more faithful to the Leninist heritage *and* more in tune with the interests of key elements within the Communist Party and Soviet proletariat—and given Stalin's continued control over the Leninist party apparatus itself—Bukharinism proved to have little more staying power than Trotskyism or Zinovievism.

Time and Soviet Labor Discipline

We have seen that the problem of situating socialism in historical time was a central problem for the Marxists of the Second International and the Bolsheviks of the 1920s. In addition, we have seen that views about when socialism might arrive in temporal terms were closely correlated in both periods with views about how labor should be organized in time. The charismatic Marxism and Leninism of Luxemburg and Trotsky argued both for the immediate achievement of socialism and for unleashing the proletariat as a continuously revolutionary force. The rational-legal Marxism and Leninism of Bernstein and Bukharin argued both for a linear evolution toward socialism and for the maintaining of the organization of the working class according to principles of rational time discipline. Finally, the neotraditional Marxism and Leninism of Kautsky and Zinoviev argued both for faith in the "dialectical" movement of history and for the socialist training of the working class by orthodox theoretical elites.

Remarkably enough, these implicit connections were made explicit in the work of important Soviet sociologists and economists of the 1920s directly engaged in the empirical study of what they believed was an emerging socialist society. In fact, the problem of how to organize socialist time discipline became a central focus of debate among party intellectuals in the NEP period. Furthermore, the three schools of thought on how to organize social time in the Soviet state were directly associated with the right, left, and center factions within the elite discussed above. The right interpretation of socialist time discipline was set out in the work of Aleksei Gastev, the left interpretation was advocated by

Platon Kerzhentsev, and the center approach—which would later be adopted by Joseph Stalin during the First Five-Year Plan—was outlined in the work of Stanislav Strumilin.[112]

The Central Institute of Labor (Tsentralnyi Institut Truda, or TsIT), founded by Gastev in 1920 with Lenin's support, was a veritable citadel of socialist Taylorism. Gastev, who had started out his revolutionary career as a "proletarian poet" composing paeans to the machine age, appeared convinced that the essence of socialism was to tie human beings ever more firmly to the temporal rhythms of machinery. "The future society," he wrote, "will be managed by special 'production complexes' wherein the will of machinism and the force of human consciousness will be fused in an unbreakable weld."[113] By adopting the methods of the "scientific organization of labor" (Nauchnaya Organizatsiia Truda, or NOT) in the context of socialism, in which the "anarchy" of capitalist production methods had been overcome, the productivity of the Soviet working population could be raised to unprecedented levels.

Gastev's mass-produced posters designed to teach Soviet proletarians "How to Work"—one of which hung in Lenin's office—set out sixteen rules of proper work behavior that could have been taken straight from Taylor's principles of scientific management.[114] His advocacy of "chronometry"—the study of work motions in minute measurements of time in order to improve the efficiency of each discrete movement by the worker—was taken up by "norm setters" in an increasing number of Soviet factories. By 1927, Gastev was explicitly comparing Karl Marx and Henry Ford as prophets who had foreseen humanity's future under a universally rational social time discipline: "Their attitudes to the working class, of course, were diametrically opposed. But it is no less curious to note something else, the thing that unites them, the thing that completely unexpectedly brings these two gigantic figures together—that is, their views on production, and their analytic approach to that production."[115] By contrast, scant attention was paid in Gastev's work to the question of how socialist time use might be made more "self-actualizing" than time use under capitalism.

Not surprisingly, workers subjected to the methods of TsIT were not as enthusiastic about Ford's contributions to modern civilization as Gastev himself. Within the party as well, there existed a widespread concern that NOT represented a new form of exploitation of the proletariat that was wholly unsuitable for a socialist state. This concern was naturally taken up by Trotsky's Left Opposition. In the summer of 1923, Trotsky personally supported the launching of an entirely different organization for the enforcement of socialist time discipline—the "Time League" created by Platon Kerzhentsev.[116]

Like Gastev, Kerzhentsev had earlier been involved in various artistic pursuits associated with Bogdanov's Proletkult movement. He had also collaborated with Gastev in the original setting up of the Central Institute of Labor. But Kerzhentsev's vision of socialist time use was quite different from that of his former colleague. Whereas Gastev emphasized the rational scientific monitoring of abstract time units, Kerzhentsev called upon those with communist consciousness to organize themselves on the basis of "spontaneous self-discipline." Under Kerzhentsev's leadership, groups of Time League enthusiasts periodically burst into the meetings of party bureaucrats and factory managers, exposing whatever wastage of time they encountered—and generally wasting quite a bit of time themselves in the process. In collaboration with the Young Pioneers, Time League propaganda was drummed into the heads of Communist Party youth, who were in this way supposed to absorb the norms of continuous revolutionary time discipline that would become universal under socialism.

The Time League at its height included some twenty-five thousand members, 40 percent of them members of the Komsomol.[117] But Kerzhentsev's charismatic approach to producing socialist time discipline was even less successful than Gastev's NOT in producing cultural transformation in Soviet Russia. As Stites aptly puts it, the League only "pitted a legion of enthusiasts and busybodies against uninterested citizens."[118] The Time League itself was quite suddenly disbanded in February 1926—not coincidentally, shortly after the Stalin-Bukharin duumvirate had consolidated its control over the party Central Committee. Gastev's organization, however, remained active—until Bukharin's defeat three years later, when the activities of TsIT were themselves sharply curtailed.[119]

Thus, the contradictions in Lenin's postrevolutionary writings between advocating Taylorism and encouraging spontaneous *subbotniki* were played out after his death in a direct competition between right and left organizations for socialist time discipline. As in Lenin's own work, however, neither side of the charisma-rationality dualism appeared capable by itself of producing distinctly socialist norms of economic time use. During the same period, however, the basis for an alternative solution to the problem of institutionalizing socialist time was worked out by an economist who would later play a central role in the designing of the First Five-Year Plan—Stanislav Strumilin.

As early as 1920, when Strumilin began working with Gastev in the TsIT, he, too, identified the problem of socialist labor discipline as central to the future prospects of the Soviet state:

> The question of the success of labor in a workers' state is a question of the purpose and meaning of its very existence. It is also a question of its life or death. Our certainty in the final victory of socialism has always rested on the fact that socialism is the most productive economic system. We firmly realize that in the historical change of economic forms the victory always goes to the more productive one. This is why after the political victory of the proletariat the central task of the worker-peasant power must be the fullest and most complete uncovering of all the economic advantages of the new order and of all the productive possibilities lying in it.[120]

But departing from the analyses of both Gastev and Kerzhentsev, Strumilin argued that these "productive possibilities" could be realized neither simply by encouraging enthusiastic production by inspired workers nor by employing a socialist form of Taylorism.

Against the belief that enthusiastic *subbotniki* alone might solve the Soviet state's economic problems, Strumilin pointed to the increasing amount of sickness, lateness, and absenteeism at work that had statistically accompanied the increased use of overtime work during the period of war communism.[121] His point was not to denigrate the heroism of the workers who had contributed to socialism in this manner, of course: "If for the salvation of the revolution we are forced to work even eighteen hours a day, the working class would of course be determined to make this sacrifice without any hesitations. But the question is how expedient this path of salvation is."[122] In fact, the author continued, an overreliance on overtime work was scientifically unsound. Strumilin's research showed that human biological limitations dictated that attempts to speed work up past a certain point would necessarily be accompanied by future losses of work productivity once laborers became exhausted:

> The human being can be regarded in the same way as any other machine. It is a living engine. Its advantage by comparison with a mechanical engine consists in its greater elasticity. It can apply itself to labors of the most diverse types in terms of the power, speed and duration that is demanded. But, of course, the general quantity of energy which it commands is highly limited and it is dispensed in

different conditions according to a coefficient of use that is far from equal in each case.

Practice has established that conditions of labor under which a person works only one third of the day at one third of his maximum force and at one third of the speed or intensity possible for him are the most advantageous from the point of view of the result. Every attempt to increase any one of these elements of the aforementioned norms entails a corresponding reduction of the two others, due to which the general result of the work decreases.[123]

At first glance, the above passage might appear to reflect an essentially abstract time orientation, according to which even revolutionary, socialist labor was stuck at a given level of productivity during a given unit of work time. Yet nothing in Strumilin's analysis contradicted the possibility of raising the "maximum force" and "intensity" attainable by workers under socialism—the standard according to which the "most advantageous" norms might then be set. Indeed, the ability more effectively to mobilize the "elasticity" of human labor energy could prove to be a crucial advantage for socialism over capitalism.

Thus, despite Strumilin's explicit opposition to the left's calls for continuous revolutionary advance on the "labor front," by 1923 he began to attack Gastev's ideal of Soviet Taylorism as well. Not so subtly disparaging Gastev's conception of the scientific organization of labor, Strumilin argued that labor productivity ultimately depended just as much on the effective use of free time for recreation as it did on the rational organization of time in the workplace—an issue to which he felt the NOTists had paid insufficient attention: "About the rationalization, or, as it is now accepted to call it, the scientific organization of labor, there has by now been much thought and discussion among us. Unfortunately, more is talked about than is done. But the question of rationalizing the *leisure* of the worker by means of the optimal utilization of his 'free' time—this question in its present form no one has yet posed."[124] The "rationalization of leisure," Strumilin argued, would allow socialist planners to ensure that Soviet citizens maintained their highest level of potential productivity—without squandering their energy on "irrational" free-time pursuits such as drinking, idling, or performing mindless household chores. In order to facilitate the effective "planning" of labor and leisure, from 1923 to 1925 Strumilin undertook a comprehensive set of "time-budget analyses" of the daily life activity of Soviet workers, peasants, and professionals, registering down to the minute the amount of time, on the average, spent by the Soviet population in "productive" and "nonproductive" activity.[125]

Thus, it was with a substantial background in the sociological study of time use in Soviet society that Strumilin began to play a central role in the State Planning Commission, Gosplan, during the formulation of a full-scale planned economy in the mid-1920s. A socialist planned economy, Strumilin argued, had to avoid both Kerzhentsev's (and Trotsky's) overreliance on enthusiasm, which ignored the scientifically ascertainable limits of the "human machine," and Gastev's (and Bukharin's) tendency to forget that socialist work time was not like capitalist work time, that the entire development of "living labor" in socialist society depended upon careful attention to the conditions under which workers conducted themselves both on and off the job. The task of scientific and revolutionary economic planners was to find an optimal balance under which the maximum revolutionary intensification of labor time would take place within rational limits.

Socialist planning, unlike capitalist planning, could thus be "teleological"—oriented toward future potentials for productivity—and need not take existing levels of economic attainment as its starting point.[126] Rather, proper planning could achieve "such a *redistribution* of the existing productive forces of society, including both labour power and the material resources of the country, as would secure to the *optimum* extent the expanded and crisis-free reproduction of these productive forces *at the most rapid possible rate*, with the aim of *maximum* satisfaction of the current needs of the working masses and of bringing them *very rapidly* to the full reconstruction of society on the principles of socialism and communism."[127] Indeed, if the right planning formula was reached, Strumilin concluded, socialism could completely outstrip capitalism: "We are bound by no laws. There are no fortresses the Bolsheviks cannot storm."[128] By 1928, this slogan had become Stalin's watchword.

5

THE

SOCIOECONOMIC

CYCLE

From Stalin to the
"Era of Stagnation"

Perhaps no question is more central to intellectual disputes about the nature of the Soviet regime than how to characterize the historical connection, or disjuncture, between Leninism and Stalinism.[1] Hanging like a shadow over every argument on either side of this issue is the incontrovertible fact that the Stalinist regime that emerged after 1929 was one of the most brutal and bloodthirsty in human history. Responsibility for the deaths of millions of peasants during the forced collectivization of the peasantry in 1932, of a large percentage of the party and state elite during the Great Terror of 1936–38, and of countless millions of inhabitants of the USSR whose lives might have been spared given proper military preparations before the Nazi attack in 1941 can be placed squarely on the shoulders of Stalin and his political associates. The reign of Stalin also saw the imposition of an all-pervasive system of spying and police terror over the population of the Soviet Union, the formation of a huge network of forced labor camps contributing no small portion of overall Soviet economic production, the deportations of entire "enemy nationalities" from their homelands, the stifling of even the slightest degree of freedom to dissent from official norms in all spheres of social life, and finally the imposition of these features of Stalinist rule on the Eastern European countries occupied by the Red Army after the defeat of Hitler.[2]

It is hardly surprising, then, given the morally repugnant nature of the Stalinist regime from a Western liberal perspective, that arguments about the nature of the link or disjuncture between Lenin and Stalin tend to be infused with a great deal of moral passion. The totalitarian paradigm, which dominated the Western analysis of this question from the 1950s until the early 1970s, was animated by a moral concern with individual freedom and an opposition to all forms of centralized administrative control over political and economic life.[3] Analysts in this school saw Stalinism as a logical outgrowth of the original dictatorial methods of party organization proposed by Lenin in 1902 and implemented by the Bolshevik elite after 1917.[4] From this perspective, attempts to separate Lenin's legacy in any way from Stalin's massacres could be seen as providing moral support for the original architects of totalitarianism—Lenin, the other Bolsheviks, and perhaps Marx and Engels as well.

The revisionists of the 1970s and 1980s, disputing what they saw as the political agenda motivating the totalitarian approach, tended to reverse this moral argument rather than attempt to separate it from the task of analysis. Revisionists attacked the totalitarian model's "continuity thesis" linking Leninism and Stalinism not only for its deemphasis of significant historical data but also because this thesis implied that the Soviet Union had been from its inception an unreformable regime of absolute repression—a notion that the revisionists saw as fueling a needless anti-Sovietism on the part of Western policy makers.[5] From this perspective, the "continuity thesis" could be seen as providing moral support for Western proponents of the Cold War and the arms race.[6]

Both the totalitarian and the revisionist schools have made crucial contributions to our understanding of the history of the establishment of Stalinism in the twenties and thirties. The totalitarian paradigm, as interpreted by sensitive historians such as Merle Fainsod, gave rise to a series of empirical and historical studies detailing the operations of political and economic institutions in the USSR that combined moral condemnation of Leninism with a healthy respect for the facts.[7] The revisionist paradigm, on the other hand, gave rise to research into topics poorly explained by the totalitarian school, such as the nature of the social support for the Bolshevik regime, the diversity of political, cultural, and economic viewpoints within and outside the party during the NEP period, and the continuing forms of social resistance to party rule even during the Stalin era.[8]

However, the different moral agendas of the two major camps in Soviet studies over the last decades have presented each with characteristic analytic problems. The totalitarian school, as Cohen points out, seemed at times to depict the entire history of the USSR since 1917 as nothing but a series of more or less clever moves by the party elite to consolidate absolute control over society.[9] The totalitarian model's stress on the historical and comparative uniqueness of the Soviet form of rule led analysts to underestimate important pressures for change operating within the regime throughout the course of Soviet history; focusing on these pressures for change in one's academic analysis of the Soviet Union was often equated with a blunting of one's moral condemnation of the communist system.

The revisionist school, by contrast, in its zeal to present the Soviet regime as subject to the same political divisions and social pressures as regimes elsewhere, often lost the ability to portray what was truly unprecedented about the Soviet revolutionary experiment. Efforts to show that despite the complexity of social and political trends over the several decades of Soviet history there was after all something peculiarly *Leninist*—let alone Marxist—about the Soviet regime were seen as necessarily providing support for the totalitarian school and, by implication, the Cold War.

The argument presented here must therefore necessarily be a controversial one from both the totalitarian and revisionist points of view. I have presented, in the preceding chapter, an interpretation of the NEP period that closely coincides with the revisionist perspective. Contrary to the totalitarian thesis, I have portrayed Lenin's plans for the building of a socialist economy as largely contradictory, if not incoherent. This incoherence, I argue, provided a great deal of room for conflicting interpretations of the revolutionary heritage; not only Trotsky and Bukharin but Zinoviev as well might well have gained control over the future course of the regime's development given different historical circumstances. There was thus no inexorable law forcing Leninist party rule to develop into the political, socioeconomic, and cultural forms of Stalinism.

However, my argument in this chapter borrows heavily from the insights of the totalitarian school, for, despite the diversity of viewpoints on economic policy in the twenties, there *is* a consistent core to the intellectual and political tradition leading from Marx to Lenin to Stalin that must not be ignored. This core, I have argued, reflects a common acceptance of the charismatic-rational conception of time—the idea that time is a force ultimately to be transcended through revolutionary action but that revolutionary action itself, in order to be effective within the realm of mundane history, must be time-disciplined. Neither pure revolutionary enthusiasm, such as that characteristic of anarchism, nor a total acceptance of historical constraints on political action, such as that characteristic of bourgeois reformism, can preserve this synthesis. Nor can a purely formal acceptance of both sides of the dualism of revolutionary time transcendence and modern time discipline qualify as true Marxism in the absence of concrete revolutionary action; this merely leads one to the establishment of a sort of "revolutionary orthodoxy" that does nothing to supplant the bourgeois order and therefore slides over into de facto reformism.

Stalin, I will argue, far from being the theoretical mediocrity he has almost universally been assumed to be, both by his political rivals and by later analysts, was the one person in the Bolshevik leadership who intuitively grasped the nature of this revolutionary time discipline at the core of Marxism and Leninism *and* who had a sense, however tentative at first, of how to create economic institutions within the USSR that would faithfully reflect this core synthesis. Stalin's political maneuvering in the 1920s was evidence not merely of his thirst for absolute power but also of the tactical flexibility afforded by an acceptance of the charismatic-rational conception of time. Thus, Stalin's decisions to ally with the center triumvirate against Trotsky and the left, then to join with Bukharin against the United Opposition of left and center, and finally to lead an attack on Bukharin and the right must not be taken merely as evidence of his unprincipled nature or of his indifference to matters of Marxist and Leninist theory. Instead, we can fruitfully compare Stalin's role in the twenties to Lenin's role during the period of the Second International, an analogous period of fragmentation and decline within the Marxist revolutionary movement. Like Lenin, who believed in a political synthesis of revolutionary action and rational discipline and who therefore could at different points support right, left, or center groups for tactical reasons without sacrificing his principles, Stalin at different times found allies among supporters of Trotsky, Zinoviev, and Bukharin while consistently distancing himself from his rivals' views on "socialist construction."

There are, of course, significant differences between Lenin and Stalin as well. Stalin, unlike Lenin in 1902, did not make his own position explicit at the outset of the struggles of the 1920s; *The Foundations of Leninism*, while providing interesting evidence of Stalin's early views, is hardly a *What Is to Be Done?* for Stalinist economics. Rather, Stalin developed his independent theoretical position slowly over the course of the 1920s, arriving at his ultimate solution to the problem of how to formulate a socialist "economy of a new type" only during the final break with Bukharin in 1928–29. This solution, encapsulated in the extremist industrialization and collectivization drives of the First Five-Year Plan, was unprecedented in the history of Bolshevik economic debate. But, as we have seen, Lenin himself had provided no consistent answer to the question of how to construct a socialist economy, and in significant respects the Stalinist economic system that emerged in the late twenties was closer to the spirit of Leninism than anything proposed by Trotsky, Zinoviev, or Bukharin. This chapter thus analyzes the socioeconomic institutions of Stalinism as a logical response—though not the only possible one—to the dilemmas of "socialist construction" under Leninist auspices.

After tracing the development of Stalin's independent theoretical views in the 1920s and analyzing the basic institutions of Stalinism as representing the concretization of a new, socioeconomic form of charismatic-rational time, I turn to an examination of another controversial question—that of the relationship between Marxism, Leninism, and the system of mass terror established by Stalin in the 1930s. While not ignoring the role of the dictator's psychological motivations here, I argue that Stalin's systematic terror must be understood, in part, as a response to the remaining dilemma of the charismatic-rational conception of time—the problem of creating a culture, and not just a socioeconomic structure, based on the mass internalization of charismatic-rational time discipline. The ultimate failure of the Soviet system to produce such a cultural transformation meant that Stalinist socioeconomic institutions would inevitably undergo relatively rapid corruption. Unable to prevent continuing cultural resistance to a socioeconomic regime of continuous revolutionary discipline in the USSR, Stalin instead arrived at a theoretical formulation that made any final resolution of the cultural problem unnecessary, while maintaining the possibility of continual revolutionary action by party cadres—namely, the thesis that "the closer we are to the final victory of socialism, the more enemies we shall have."

Finally, this chapter concludes with a brief examination of the period of decline of the socioeconomic cycle of Leninist development from 1953 to 1985. After Stalin's death, I argue, the same process of breakdown and fragmentation of the revolutionary movement that we have seen in the previous two cycles of development took place once again.

Having decisively rejected Stalin's "solution" to the cultural dilemma of the charismatic-rational conception of time—mass terror—the post-Stalin leadership endeavored to find some other path toward communism while preserving the theoretical core of Marxism, the political primacy of Leninism, and the major socioeconomic institutions of Stalinism. But, as in the period after the deaths of Marx and Engels in the theoretical cycle, and as in the years after the death of Lenin in the political cycle, the decades after Stalin's death saw a split between left, right, and center approaches to this problem. The rightist evolutionary approach to creating charismatic-rational culture was represented by Malenkov, the leftist revolutionary approach by Khrushchev, and the centrist, orthodox approach by Brezhnev. As in the previous two periods of decline in the Marxist-Leninist revolutionary movement, none of these three approaches proved capable of successfully resolving the cultural problem, and the result was stagnation, fragmentation, and the potential collapse of the Soviet regime.

Throughout this chapter, I do my best to remain faithful to the ideal of a value-neutral analytic perspective. Despite the moral outrage one might justifiably feel toward the Stalinist system and the man who presided over it—a sense of outrage that I personally share—a consistent Weberian social scientist cannot allow personal values to color the analysis of the social and political factors leading to Stalinism's historical emergence. It is not that Weberian analysis presumes a kind of "cultural relativism" that declares all moral standards to be invalid a priori. Rather, it proceeds from an insistence that *analysis* and *judgment* are two conceptually separable activities. Stalinism, like any other political or socioeconomic system, must be approached historically and comparatively, not in order to deny its evil nature, but in order to situate even the cruelest forms of *Herrschaft* in a framework allowing for interpretive understanding.

Stalin as Theorist, 1924–1928

The subject of Joseph Stalin's rise to power in the 1920s has been explored by countless writers. But, despite the sheer volume of scholarly and literary material on the topic, Stalin's positions on questions of Marxist and Leninist theory in the debates of the NEP period have rarely been examined as important on their own terms. The standard account of Stalin's role in the twenties stresses his skillful manipulation of the instruments of political power, especially his ability as general secretary to stack the Bolshevik Party with his own supporters, and sees ideological debates as at best a secondary aspect of the power struggle.[10] Stalin is presented, rather paradoxically, as a completely unoriginal thinker who simply happened to direct radically original policies after 1928.

Such a verdict is, remarkably, common to analysts of the totalitarian school, Trotskyists, and revisionists alike. According to Fainsod, for example, Stalin's appeals to Leninism were nothing more than a mask for his drive for totalitarian control: "Stripped of its propaganda verbiage, the Stalinist program foreshadowed a profound extension of the scope of totalitarian power. The peasantry was to be brought to heel and tied to state ends. The surpluses extracted from the peasantry were to provide the wherewithal to create a powerful industrial structure which would render the Soviet citadel impregnable."[11] E. H. Carr, whose interpretation of early Soviet history was profoundly influenced by Trotsky, agrees that Stalinism had little to do with the theoretical aspects of Marxism or Leninism: "What Stalin brought to Soviet policy was not originality in conception, but vigor and ruthlessness in execution. When he rose to power in the middle nineteen-twenties, he became, and was determined to remain, the great executor of revolutionary policy. But the course of events makes it clear that he had at that time no vision of where that policy would lead."[12] Finally, we have the appraisal of Stephen Cohen, who, despite his disagreement with practically every other element of the Trotskyist and totalitarian interpretations of early Soviet history, nonetheless agrees that Stalin's rule was based on "brute machine power" and that he "had no standing whatsoever" as a man of theory.[13] Given the range of divergence among the authors presented here on most issues, this is a remarkable scholarly consensus.[14]

It is undeniably the case that Stalin's theorizing lacks the stylistic polish and subtlety of the works of his main competitors for power in the 1920s. Trained in the catechisms of the Tiflis Theological Seminary, Stalin lacked his rivals' long years of exposure to Western intellectual trends; his prose strikes readers accustomed to a more erudite rhetorical style as crude and unpersuasive—which is how it struck Trotsky, Bukharin, and many other Old Bolsheviks. But it does not follow from this that Stalin's theoretical position was, as Carr puts it, "intellectually and emotionally trivial."[15] In fact, if we choose to concentrate on issues not of style but of substance, Stalin's argumentation begins to manifest a ruthless logical consistency. It is a strange sort of logical consistency, to be sure, built upon an absolutized conception of society as a realm of continuous class warfare—but it is a consistency that remains firmly within the theoretical and political discourse established by Marx and Lenin.

In fact, Stalin's position in the debates of the 1920s is, upon closer examination, strikingly analogous to Lenin's position vis-à-vis the theorists of the Second International. First of all, like Lenin, Stalin adopted a position explicitly opposed to both right and left tendencies within the revolutionary movement, ultimately treating them as common expressions of a bourgeois subservience to "spontaneity." Also like Lenin, Stalin originally allied himself with the center, orthodox position within the theoretical debate, as a member of Zinoviev's triumvirate. But just as Lenin had ultimately broken with Kautsky over the issue of the appropriateness of *political* revolution in Russia, Stalin broke with Zinoviev over the issue of the possibility of *socioeconomic* "socialism in one country." Finally, like Lenin, Stalin discovered a concrete institutional answer to the question of how to combine revolution and discipline in everyday practice: the creation of a Stalinist economic system of "planned heroism" paralleling the political institutionalization of Marxism in Lenin's "party of professional revolutionaries." I now turn to examine the development of Stalin's independent theoretical position in more detail.

Stalin's position as an orthodox Leninist, resisting left and right "deviations" alike, was in his early career expressed primarily through his nearly complete fidelity to Lenin himself on matters of principle. Stalin's first major theoretical work, *Marxism and the National Question*, was written under Lenin's sponsorship and close supervision; it amounts to a rather workmanlike restatement of Lenin's various pronouncements on issues of nationality, collected in one comprehensive volume.[16] From that point until shortly before Lenin's death, Stalin avoided any open theoretical dispute with Lenin, despite their occasional differences on debates about revolutionary tactics. This reticence, which reflects not only Stalin's political savvy but also his genuine admiration for Lenin's theoretical genius, stood the Georgian in good stead in the battles of the twenties. Both Trotsky and Bukharin had often engaged Lenin in open, serious intellectual debate; Zinoviev had been more circumspect, but in 1917 he, along with Kamenev, had openly betrayed Lenin's plans for the October Revolution—a black mark on his record that was especially damaging to a would-be defender of Leninist orthodoxy.

Stalin, of course, had his own special problem in claiming fidelity to Lenin's legacy: the fact that the dying leader had explicitly urged the party to remove Stalin as general secretary because of his "rudeness" and his dubious ability to use his enormous bureaucratic power with "sufficient caution."[17] In addition, Stalin had clashed violently with Lenin over the brutal means he had employed in his position as commissar of nationalities in suppressing the autonomy of Soviet Georgia. But in a revolutionary movement that placed such an emphasis on theoretical purity, these *policy* disputes were, ironically, in some ways easier for Stalin to live down than were the *doctrinal* disputes with Lenin of his rivals. Stalin could, and did, insist that his differences with Lenin were only tactical ones, and even that his "rudeness" could be presented as a tactical asset; as he confessed at a Central Committee Plenum meeting in October 1927: "Yes, I am rude, comrades, towards those who rudely and treacherously wreck and split the party."[18] His opponents, on the other hand, had come out openly against key components of Lenin's theoretical work itself—a potentially much more damning indictment. That Stalin's rivals themselves saw this problem in a similar light is demonstrated by their evident reluctance, until it was much too late, to make Lenin's testament public. Lenin's criticisms, in the same document, of Trotsky's "non-Bolshevism" and Bukharin's incompletely Marxist theoretical views must have seemed in some ways worse accusations than his purely political complaint about Stalin's excessive power. Knowing this, Stalin confidently presented himself in the party struggles of the twenties as Lenin's most dependable and consistent ally.

However, Stalin's basic agreement with Lenin's theoretical views was based on more than political posturing. After Lenin's death, Stalin's arguments about the nature of socialist construction in the USSR continued to manifest an intuitive distrust of left and right "deviations" from orthodoxy, combined with a keen understanding of the need for disciplined revolutionary advance, that mark Stalin as a true believer in the charismatic-rational conception of time. Already by the end of 1924, we see Stalin developing the outlines of his distinctive conception of how to weld together charismatic time transcendence with rational, modern time discipline—and not, like Zinoviev, simply to do so through protestations of fidelity to "Leninism." On the contrary, alone among all his rivals in the 1920s, Stalin grasped the central necessity of building a charismatic-rational "economy of a new type" in the Soviet Union if the momentum of the Marxist revolutionary movement was to be maintained.

Stalin's first independent theoretical work, *The Foundations of Leninism*, based on lectures given at the Sverdlov University and published in *Pravda* in April–May 1924, is in this respect worth examining in some detail.[19] The main thrust of this essay is an almost liturgical assertion of Leninist orthodoxy. Stalin's constant use of a rhetorical question-and-answer style, his reliance on phrases such as "it scarcely needs proof" at controversial points in the argument, and his heavy-handed dismissal of alternative interpretations have usually been taken as signs of Stalin's intellectual weakness—thus Deutscher's verdict that "what [Stalin] had to say on the subject [of Leninism] was so unoriginal and dull that it hardly deserves to be summarized."[20] But to dismiss Stalin's *Foundations of Leninism* as merely an ill-advised foray into matters of theory by a man greedy for recognition from his more erudite peers would be to miss an important aspect of the future dictator's rise to power, for this pamphlet contains many of the theoretical seeds that would later sprout into the full-blown theory and practice of Stalinism.

To begin with, we must bear in mind that the proper form of acknowledging one's debt to Lenin was itself a charged theoretical and political issue in the Bolshevik Party after the founder's death. In this light, Stalin's assertion of a quasi-religious Leninist orthodoxy in 1924 was itself part of his independent theoretical stance. While both Trotsky and Bukharin were attempting at this time to justify their disagreements with Lenin at particular points, Stalin took a different tack, arguing implicitly that no departure from the words of Lenin could ever be true to revolutionary Marxism in the present historical era. "Leninism *is* Marxism in the era of imperialism and the proletarian revolution," he argued,[21] thereby undercutting the possibility of appealing to aspects of pre-1917 Marxism in order to modify "Leninism" in any way. This may strike us as a rather restrictive and anti-intellectual approach to theoretical matters, and we may prefer the more creative interpretations of Lenin's legacy given by Trotsky and Bukharin. But it is worth pointing out that Stalin's absolutization of "Leninism" as dogma was hardly without precedent in the history of Marxism. Indeed, this was precisely how Lenin, following Kautsky and Engels, himself interpreted "Marxism"—as an integral "revolutionary theory" without which a revolutionary movement was impossible.

In essence, Stalin sided with Zinoviev, his partner in the triumvirate, on the question of whether Leninism should be considered a flexible heritage or a quasi-religious doctrine. But Stalin parted company with Zinoviev on a further question, namely, how to interpret the various inconsistencies and vacillations in Lenin's speeches and writings after the revolution. Zinoviev, it appears, took Lenin's contradictory pronouncements on socioeconomic policy to be as central to "Leninism" as

Lenin's political views. Thus his economic strategy, like Lenin's, remained fundamentally incoherent.

Stalin, on the other hand, seemed intuitively to grasp that there was a crucial consistent *core* to Lenin's position, a core that remained constant despite Lenin's inconsistent economic policy after 1917. That core was the conception of the "professional revolutionary" party laid out in *What Is to Be Done?* in 1902—the political synthesis of charisma and discipline that distinguished Lenin's views from those of all his theoretical rivals in the Second International. Lenin's various pronouncements on socioeconomic matters, in this light, could be taken as only *relatively* binding—appropriate for particular historical periods—whereas the conception of Bolshevik revolutionary discipline was to be considered *absolutely* binding. Stalin seemed to realize that Lenin's solution to the problem of how to institutionalize Marxism politically was by itself insufficient as a solution to the problem of how to institutionalize Marxism economically. It would be Stalin's goal to *extend* the core principles of political Leninism to the workings of the Soviet economy—something Lenin himself had failed to accomplish.

Stalin's division of "Leninism" into two parts—a core consisting of Lenin's views on the nature of the party, and a less central series of tactical pronouncements on economic issues—lies at the basis of the presentation of Leninist orthodoxy given in *The Foundations of Leninism*. By far the greatest weight in the pamphlet is given to issues of party unity and discipline. Most of the section subtitled "Theory," for example, consists of a restatement of the strictures against "spontaneity" in the revolutionary movement set out in *What Is to Be Done?* In the penultimate section entitled "The Party," Stalin stresses the need for "iron discipline" throughout the period of the construction of socialism *after* the revolutionary seizure of political power as well: "The proletariat needs the Party not only to achieve the dictatorship; it needs it still more to maintain the dictatorship, to consolidate and expand it in order to achieve the complete victory of socialism."[22] Truly iron discipline, Stalin continues, requires a ruthless struggle against the appearance of factions in the party, owing to the pernicious influence of "opportunist elements":

> The source of factionalism in the Party is its opportunist elements. The proletariat is not an isolated class. It is constantly replenished by the influx of peasants, petty bourgeois and intellectuals proletarianized by the development of capitalism. . . .

In one way or another, all these petty-bourgeois groups penetrate into the Party and introduce into it the spirit of hesitancy and opportunism, the spirit of demoralization and uncertainty. It is they, principally, that constitute the source of factionalism and disintegration, the source of disorganization and disruption of the Party from within.[23]

For this reason, Stalin concludes, "the Party becomes strong by purging itself of opportunist elements."[24]

All of this, it must be noted, is no more than a restatement of Lenin's own views on this subject. Indeed, Stalin backs himself up throughout this section of *The Foundations of Leninism* with quotes from Lenin that express very much the same thing—although perhaps one could argue that Stalin's formulations are somewhat more sweeping and generalized than Lenin's. Essentially, however, Stalin remained faithful to Lenin's view that "revolutionary discipline" within the party was the highest political principle.

This in itself does not, in fact, distinguish Stalin very much from the other major Bolshevik leaders of the period, who, as I have argued previously, also held to the principle of party discipline and unity even when this fundamentally undermined their attempts to oppose the policies of the party leadership. What is striking about *The Foundations of Leninism* is not its stress on party discipline but rather the way this stress contrasts with Stalin's marked deemphasis of Lenin's economic policies, and in particular, of the NEP. Unlike Zinoviev, Trotsky, or Bukharin, Stalin spent little time attempting to integrate Lenin's views on socioeconomic matters into his general presentation of "Leninism."

To be sure, in his discussion in a section entitled "The Peasant Question," Stalin does present the basic elements of the NEP as Lenin outlined it. But his lack of enthusiasm for the NEP is clear from the striking fact that he refuses to mention it by name anywhere in this section. In addition, Stalin clearly stresses the transitory nature of the NEP, claiming that its major components merely represented "how Lenin outlined the immediate tasks of economic construction on the way to building the foundations of socialist economy."[25] Certainly, this tentative formulation lacks the decisive tone of Stalin's pronouncements on the party. More tellingly, when Stalin finally does mention the NEP by name, it is in the section entitled "Strategy and Tactics"; here Stalin characterizes the revival of private trade under the NEP as a "tactic" comparable in significance to Lenin's prerevolutionary proposal to create an all-Russian underground newspaper. The obvious implication here was that the NEP was no more to be considered an eternal component of Leninism than the original *Iskra*—which Lenin began to attack as soon as its editorial board rejected his political line! Finally, in discussing the question of whether Lenin, in adopting the NEP, had essentially advocated "reform" *over* "revolution," Stalin resolutely denies any such thing. The contrast with Bukharin here is striking:

> Under certain conditions, in a certain situation, the proletarian power may find itself compelled temporarily to leave the path of the revolutionary reconstruction of the existing order of things and to take the path of its gradual transformation. . . . It cannot be denied that in a sense this is a "reformist" path. But it must be borne in mind that there is a fundamental distinction here, which consists in the fact that in this case the reform emanates from the proletarian power, it strengthens the proletarian power, it procures for it a necessary respite, its purpose is *to disintegrate, not the revolution, but the nonproletarian classes.*[26]

A clearer statement as to Stalin's ultimate attitude toward the NEP—and as to his ultimate intentions regarding the peasantry and other "non-proletarian classes"—can hardly be imagined.

The Foundations of Leninism thus establishes Stalin as a defender of the orthodox interpretation of Lenin's politics of revolutionary discipline while placing him in decided opposition to those who would give Lenin's economic policies the same venerated status. This, however, leads directly to a further issue: whether or not Stalin *himself* had any notion of how to build socioeconomic institutions in the USSR that would faithfully reflect what he saw as the crucial core of Leninism—what we have analyzed as the political synthesis of charismatic and modern time orientations. In fact, Stalin does begin to address this question in *The Foundations of Leninism*. In the section of the pamphlet dealing with the party, Stalin argues that the primary task of the victorious revolutionary party is to prepare the grounds for a socialist economy by inculcating discipline on a mass scale:

> Now what does to "maintain" and "expand" the dictatorship mean? It means imbuing the millions of proletarians with the spirit of discipline and organization; it means creating among the proletarian masses a cementing force and a bulwark against the corrosive influences of the petty-bourgeois elemental forces and petty-bourgeois habits; it means enhancing the organizing work of the proletarians

in re-educating and remolding the petty-bourgeois strata; it means helping the masses of the proletarians to educate themselves as a force capable of abolishing classes and of preparing the conditions for the organization of socialist production.[27]

But this passage leaves open the question of just what sort of "discipline" would be necessary for the "organization of socialist production." Though there is no evidence that Stalin reached any final solution to this problem before 1928, already in this early work Stalin appears to be groping toward what would become the distinctive feature of Stalinist economics—the idea of a specifically Leninist "style in work" based on a combination of "Russian revolutionary sweep" and "American efficiency."[28]

The concept of a Leninist "style in work" combining revolutionary action and efficiency has many antecedents in Lenin's own writings. Most directly, it is derived from Lenin's notion of the "professional revolutionary" activity of Bolshevik party cadres—activity both time-disciplined and oriented toward the transcendence of time. In addition, as we have seen, Lenin urged at different times both revolutionary enthusiasm for work and increased attention to questions of efficiency. But the notion of a *simultaneous combination of revolutionary and time-disciplined orientations in everyday work habits* as the key to creating a socialist economy was a theoretical synthesis that was Stalin's own.

That this section of Stalin's pamphlet represented a first attempt by the general secretary to add something of his own to the Leninist canon seems likely for three reasons. First, the placement of this section at the end of the work—seemingly incongruously following the section on "the party"—suggests that Stalin wanted to give it special emphasis. Secondly, this section has only two short quotations from Lenin, neither of which directly discusses the concept of "style in work." Instead, Stalin's major supporting materials here are two contemporary novels satirizing communist work habits—the only time any sources other than Marx or Lenin are utilized in the entire work. Finally, the very brevity of this section—it is only two pages long—suggests that Stalin knew he was going out on a theoretical limb; since the main purpose of *The Foundations of Leninism* was to assert revolutionary party discipline as the core of Leninist orthodoxy, Stalin perhaps felt it prudent to give here only a brief intimation of his future extension of purely political Leninism into the realm of socioeconomic organization.

In summary, *The Foundations of Leninism*, rather than demonstrating Stalin's theoretical incompetence, lays out a remarkably consistent and original interpretation of Lenin's legacy. By treating "Leninism" as a sacred orthodoxy, by defining that orthodoxy as centered on the core conception of revolutionary party discipline, and by treating Lenin's postrevolutionary economic policies as mere tactical adjustments to shifting circumstances, Stalin set himself apart from every other Bolshevik leader of the period. He did so, furthermore, in much the same way that Lenin set himself apart from all the warring factions of the Second International. Finally, in concluding his pamphlet with a call for the creation of a "special type of Leninist worker" who would combine American efficiency with Russian revolutionary sweep, Stalin laid the foundations for what would later become his most distinctive contribution to the Marxist revolutionary tradition—the institutionalization of an economy based on "planned heroism," one that would synthesize charismatic and rational time orientations, for the first time in history, in everyday economic life.[29]

However, *The Foundations of Leninism* did leave unresolved certain crucial theoretical questions concerning the nature of socialist construction. Chief among these was the question of the feasibility of building socialism in a backward, overwhelmingly peasant country such as Russia without the help of victorious revolutionaries in the more advanced capitalist states of Europe. This was, of course, the same question faced by Lenin after the publication of *What Is to Be Done?*; Lenin's worked-out theoretical defense of the idea of "revolution in one country" came only after his break with Kautsky, with the publication of *Imperialism*. Remarkably, within six months of the publication of *The Foundations of Leninism*, Stalin had developed an analogous defense of the possibility of creating a socialist economy in the USSR without external revolutionary advance: the conception of "socialism in one country."[30]

The Foundations of Leninism, I have argued, already marked Stalin as an orthodox Leninist decisively opposed to both left and right interpretations of Lenin's legacy. According to Stalin, the only proper understanding of Leninism was as a doctrine of disciplined revolutionary class struggle led by a vanguard party; a single-minded embrace of either pure revolutionism or pure evolutionism thus stood outside the boundaries of the doctrine. However, Stalin's departure from the center form of Leninism represented by Zinoviev was not yet marked. Although important differences with Zinovievian orthodoxy were implicit in Stalin's work, they were not immediately apparent in May 1924. This is, of course, logical given Stalin's alliance with the other triumvirs, Zinoviev and Kamenev, against Trotsky and the left, throughout the year. Indeed, if one accepts the conventional account of Stalin as a theoretical nonentity interested only in expanding his own power, it is not at all clear why he would have chosen at this stage to break with Zinoviev and Kamenev at all. The logical course of action for a man in Stalin's position, assuming he had no independent theoretical views, would have been simply to continue to amass power through his position as general secretary, leaving Zinoviev in a position of nominal theoretical leadership. Over time, Zinoviev's independent power base in the Leningrad party organization would have eroded anyway, given Stalin's power of appointment at the lower levels of the party; indeed, this had already happened to a significant extent by 1925, as Stalin's relatively rapid subordination of the Leningrad opposition later that year demonstrated.

From this perspective, Stalin's decision to launch the new and extremely controversial theoretical principle "socialism in one country" at the end of 1924 requires explanation. Why did Stalin choose to embrace a position which was sure not only to lead to a break with the "internationalist" Zinoviev, but ultimately ran the risk of giving Zinoviev and Trotsky common ground for opposition, despite the previous bitter antagonism between these two men?

In fact, Stalin's break with Zinoviev in 1925 over the issue of the feasibility of "socialism in one country" marked a crucial phase in the former's emergence as an independent Leninist theorist opposed to the long-term continuation of NEP economic orthodoxy—a phase that has been obscured by the usual tendency of Western analysts to concentrate exclusively on Trotsky's and Bukharin's oppositions in the 1920s. The main thrust of "socialism in one country" was actually an attack on Zinoviev, who stood for the defense of the status quo at home and revolutionary advance abroad. Stalin's assertion of the possibility of building "socialism in one country" effectively reversed this equation, stressing the need for *disciplined revolutionary action* in the domestic sphere over what amounted to a new version of Kautsky's disciplined "revolutionary waiting" for world revolution.

The parallel with Lenin's break with Kautsky here is quite striking. Lenin, after all, broke with his erstwhile ally Kautsky over the question of whether one could legitimately make a political revolution in backward Russia without waiting for the productive forces to mature and without waiting for proletarian revolution in the advanced capitalist countries. Now Stalin broke with Zinoviev over the legitimacy of making a socioeconomic revolution independently of the European proletariat. Stalin's attack on Zinoviev's position as a symbol of international doctrinal orthodoxy as head of the Comintern, too, closely resembled Lenin's attack on Kautsky's position as chief theorist of the Second International. Finally, just as Lenin's attack on Kautsky preceded Lenin's decision two years later to launch the Bolshevik revolution, Stalin's attack on Zinoviev prepared the ground for the general secretary's launching of his "third revolution" in 1928–29.

Thus, by the end of 1924, Stalin had developed many of the major elements of his independent theoretical position on the nature of Leninism and the path toward "socialist construction" in the USSR. These elements were as follows: first, an emphasis on an orthodox interpretation of Lenin's views of "professional revolutionary" discipline within the party; second, a deemphasis of Lenin's pronouncements on economic matters, which were presented as purely tactical; and finally, an assertion of the feasibility and legitimacy of building "socialism in one country" given revolutionary advance on the socioeconomic front in the near future. Despite Stalin's continual shifting of alliances in the twenties, he never departed fundamentally from any of these positions.

From this perspective it becomes clear that Stalin never became wholly reconciled to the theoretical views of any of his temporary allies in the struggle for power after Lenin's death. Rather, like Lenin, he utilized in a tactical way any available support in these struggles, without compromising on what he considered to be matters of principle. Stalin could agree with Zinoviev that Trotsky had violated the essential core of Leninism by attacking the "bureaucratic" nature of party rule in 1923, and so he heartily joined in Zinoviev's condemnation of Trotsky as a "leftist" who had underestimated the revolutionary potential of the Russian peasantry. But for the time being, he refrained from any departure from orthodoxy on the question of whether socialism could be completely built without the aid of world revolution.[31] With Trotsky utterly defeated at the end of 1924, he revealed his position as an advocate of building "socialism in one country"—and welcomed Bukharin's support for this, despite the fact that Bukharin's rightist, evolutionary interpretation of the slogan had very little in common with Stalin's views. In the short run, however, Stalin's embrace of the NEP as a tactical policy coincided perfectly with Bukharin's advocacy of NEP on a doctrinal level; there was therefore nothing hypocritical in Stalin's continual statements from 1925 to 1927 about the importance of maintaining the "link" between the working class and the peasantry—for the time being.

Finally, however, with the destruction of the United Opposition of

Trotsky and Zinoviev in 1927, Stalin began to unveil his full program. To anyone who had followed Stalin's statements closely, and who took his theoretical pronouncements seriously, the radical nature of that program should not have come as a complete surprise. As Tucker notes, as early as November 1925, Stalin was comparing the situation in the country to the days immediately preceding October 1917 and was calling for "Leninist hardness" in the upcoming struggle against capitalist elements in the economy. In December 1926, he asserted before the Comintern Executive Committee: "Building socialism in the USSR means overcoming our own Soviet bourgeoisie in the course of a struggle."[32] Already he sounded very much like the economic revolutionary he was later to become in practice:

> We have won the dictatorship of the proletariat and have thereby created the *political* basis for the advance to socialism. Can we by our own efforts create the *economic* basis of socialism, the new economic foundation necessary for the building of socialism? What is the economic essence and economic basis of socialism? Is it the establishment of a "paradise" on earth and universal abundance? No, that is the philistine, petty-bourgeois idea of the economic essence of socialism. To create the economic basis of socialism means welding agriculture and socialist industry into one integral economy, subordinating agriculture to the leadership of socialist industry.[33]

Following Lenin's example, Stalin was preparing to overcome the gap between Marxist theory and Soviet practice that had emerged during the NEP period in a practical way: by launching a new revolution. After 1928, he did precisely that.

Understanding Stalin's independent theoretical stance in the 1920s forces us to reevaluate the nature of Stalin's tactical maneuvering during the period, for the myth of Stalin's lack of theoretical ability has presented a skewed picture of the future dictator's rise to power. One cannot, of course, deemphasize the importance of Stalin's megalomania and of his control over the machinery of the party in accounting for his eventual victory over his rivals. No doubt, these were necessary conditions for Stalin's political success. But such factors are not a sufficient explanation for Stalin's emergence as a socioeconomic revolutionary who subjected his country to unprecedented violence and upheaval and who actually created a fundamentally new type of socioeconomic order. For those who followed Stalin along this path, he was not merely an aspiring dictator, nor could he have been seen as a theoretical mediocrity. For those who responded to Stalin's call to begin the "third revolution" in 1928–29, Stalin was a man of vision.

The "Third Revolution," 1928–1932

Stalin's all-out drive for the collectivization and industrialization of the USSR during the First Five-Year Plan was the most consequential period of Soviet history. The events of these years essentially determined the basic patterns of Soviet economic activity until the final collapse of the Soviet system under Gorbachev. Stalin's belief that his policies were the historical equivalent of Lenin's revolutionary seizure of power in 1917, a "third revolution" that would create the foundations of a new type of economy in the Soviet Union, was in this respect quite justified. However one judges the outcome from a moral perspective, there is no question that the period of the First Five-Year Plan witnessed cataclysmic social change of a revolutionary nature.

At the same time, however, Stalin's economic policies also represented a solution to the puzzle of how to institutionalize socioeconomic activity while maintaining a charismatic-rational orientation toward time. As we have seen, the revolutionary and time-disciplined components of Marxism, successfully synthesized in political terms in Lenin's "party of professional revolutionaries," had become unglued in Bolshevik economic theory and practice after 1917. Lenin himself had alternated praise for the spontaneous enthusiasm of workers on voluntary *subbotniki* with calls for the strictest enforcement of time discipline on the model of Taylor and Ford, with no apparent sense of the contradiction involved therein. And Lenin's last writings seemed only to add to the confusion.

After Lenin's incapacitation, the left, led by Trotsky, had attempted to resolve the contradiction through a reliance on party-led revolutionary advance on all fronts, both foreign and domestic. But domestically, utilizing the enthusiasm of the Soviet working class for the construction of socialism quickly ran up against obvious limits in an overwhelmingly nonproletarian country. Ultimately, therefore, Trotsky's economic analysis held out little, if any, hope for socialist construction in the USSR without the support of the revolutionary proletariat of Europe. The right, led by Bukharin, seeing the same empirical obstacles to a working-class–based socialist economy in peasant Russia, instead based its strategy on an open embrace of trade with the peasant masses as the foundation of future economic growth. But Bukharin's policies left no room whatsoever for revolutionary enthusiasm in economic activity. Both on the macroeconomic level, where slow evolution was promoted over rapid breakthroughs in production, and on the microeconomic level, where "rationalization" campaigns in Soviet factories seemed increasingly indistinguishable from the exploitation of proletarian labor time under capitalism, any access to the charismatic, timeless realm was blocked. In short, as had been the case with Luxemburg and Bernstein, the left presented only the possibility of revolutionary failure, while the right presented only the possibility of bourgeois success.

As a result, the Soviet economy during the NEP remained an unstable amalgam of "socialist" and "capitalist" elements. This was, after all, Lenin's contradictory legacy in socioeconomic affairs, a legacy more or less faithfully continued by Zinoviev until 1925. Then, with the adoption of Bukharin's evolutionary economic policies from 1925 to 1927, the fragile balance between the party's revolutionary orientation and its economic gradualism rapidly tilted in favor of the latter. By 1928, the plausibility of the Bolshevik Party's claim to legitimacy as a force capable of mastering time itself was seriously threatened.

None of this made Bukharin's defeat and the destruction of the NEP inevitable. But the long-term continuation of a Bukharinist form of market socialism would hardly have seemed like a triumph to Marx or Lenin. After all, the whole point of overthrowing capitalism, and of creating a disciplined revolutionary party capable of leading the attack, had been to usher in a world beyond the time constraints operating in class-based society, beyond the alienating subjection of the creative force of human labor to clock discipline for the sole purpose of expropriating surplus value to line the pockets of the ruling class. The proletarian revolution could not be considered complete, then, until control over time was somehow restored to those participating in everyday productive activity. At the same time, however, as long as world revolution and full communism remained a distant hope, time discipline had to be maintained at all costs, lest the tender shoots of the future society be crushed by the continuing power of capitalism. In short, the problem facing the Bolsheviks in the late 1920s was not simply how to industrialize a peasant country but how to do so in a Marxist and Leninist manner. And this required the creation of socioeconomic institutions based, like Leninist political institutions, on a synthesis of charismatic time transcendence and rational time discipline.

Stalin's "third revolution," despite its enormous costs in the short and long term for the Soviet population, was a logical response to this problem. In the years of the First Five-Year Plan, Stalin presided over the creation of a fundamentally novel type of socioeconomic order, one based on the institutionalization of a system of "planned heroism." Following up his earlier call for the creation of a "special type of Leninist worker" who would combine "American efficiency" with "Russian revolutionary sweep," and concretizing the intellectual insights of Gosplan economists such as Strumilin, Stalin's planning system made charismatic-rational time a practical principle for millions of Soviet workers and managers.

Like Lenin's proposal for a "party of a new type" in *What Is to Be Done?*, Stalin's "economy of a new type" contained three basic elements: first, an attack on both left and right positions on the peasant question as essentially equivalent forms of "bowing to spontaneity"—spontaneity that would be overcome through an all-out drive for the collectivization of agriculture; second, the conception of Five-Year Plans that should, in principle, be completed ahead of time; and third, a system of plan targets and work norms that were to be *overfulfilled* rather than merely fulfilled. I will examine each of these points in turn.

The collectivization of agriculture was Stalin's most radical departure from Soviet socioeconomic policies of the twenties. There was no explicit precedent in Bolshevik debate for this full-scale attack on the way of life of hundreds of millions of peasants. The deaths of millions of the most energetic and skilled members of the rural population in the drive to "liquidate the kulaks as a class" and in the ensuing famine of 1932–33 dealt the Soviet countryside a blow from which it may never fully recover.[34] And yet the forced collectivization of agriculture followed quite logically from the interpretation of Leninism articulated by Stalin from 1924 on. Believing as he did in a *revolutionary building of socialism in one country* opposed equally to Trotsky's charismatic revolutionary internationalism, Zinoviev's orthodox internationalism, and Bukharin's rational-legal evolutionary path to socialism, Stalin had stated early on that the NEP period of reformism was only a necessary respite to prepare the party to "liquidate . . . the non-proletarian classes." From this perspective, the extension of the NEP beyond the point necessary for the consolidation of the revolutionary movement was not only tactically wrong, but it threatened the destruction of the revolution itself.

Stalin's argument on this score closely paralleled Lenin's argument against the "Economists" and "terrorists" in *What Is to Be Done?* and indeed was undoubtedly inspired by it. As for Lenin, the key problem for Stalin was the problem of combating "spontaneity"—in this case, symbolized by the need of the party under NEP conditions to respond to peasant interests rather than directing economic activity in a socialist manner. As the grain crisis of 1927–28 seemed to demonstrate, the party that had made a political break with bourgeois time in 1917 was still hostage to the deeper hold of the temporal cycles of agricultural production. Citing Lenin's statement that "small production *engenders* capitalism and the bourgeoisie continuously, daily, hourly, spontaneously, and on a mass scale," Stalin in October 1928 argued that a continued pro-peasant policy would now mean the restoration of capitalism in the USSR: "It is clear that, since small production bears a mass, and even a predominant character in our country, and since it *engenders* capitalism and the bourgeoisie continuously and on a mass scale, particularly under the conditions of the NEP, we have in our country conditions which make the restoration of capitalism *possible*."[35] Stalin drew the logical conclusion from this in Leninist terms: "It follows that we cannot build socialism in industry alone and leave agriculture to the mercy of spontaneous development on the assumption that the countryside will automatically follow the lead of the towns."[36] Finally, paralleling Lenin's analysis of the tendencies within the Russian Marxist movement in 1902, Stalin asserted that the responses of both left and right to the "spontaneous" generation of capitalism in the countryside were essentially equivalent, since neither held out any hope for transcending bourgeois constraints on economic activity:

> Where does the danger of the *Right*, frankly opportunist, deviation in our Party lie? In the fact that it *underestimates* the strength of our enemies, the strength of capitalism: it does not understand the mechanism of the class struggle under the dictatorship of the proletariat and therefore so readily agrees to make concessions to capitalism, demanding a slowing down of the rate of development of our industry. . . .
>
> Where does the danger of the "*Left*" (Trotskyist) deviation in our Party lie? In the fact that it *overestimates* the strength of our enemies, the strength of capitalism; it sees only the possibility of the restoration of capitalism, but cannot see the possibility of building socialism by the efforts of our country; it gives way to despair and is obliged to console itself with chatter about the Thermidor tendencies in our Party. . . .
>
> You see, therefore, that both these dangers, the "Left" and the Right . . . lead to the same result, although from different directions.
>
> Which of these dangers is worse? In my opinion one is as bad as the other.[37]

Neither left nor right, then, could overcome the force of "spontaneity" in Soviet economic life. The right simply bowed to the spontaneity of empirical economic trends that could only reproduce bourgeois consciousness on a mass scale; the left bowed to the spontaneity of despairing intellectuals who had no patience for a disciplined struggle against the forces of capitalism on the domestic front.

Identifying the existence of private peasant agriculture as the source of capitalism's continual, spontaneous generation in the USSR, and rejecting both left and right responses to the grain crisis of 1927–28 as essentially equivalent, bourgeois policies, Stalin instead advocated the strict subordination of agriculture to the "socialist sector" of the economy—much as Lenin had advocated the subordination of the spontaneous workers' movement to party discipline in *What Is to Be Done?* Stalin's collectivization drive was thus first and foremost a battle against the power of spontaneity over the Soviet economy, the necessary precondition for the construction of a truly socialist economy in the Soviet Union that might prove capable of beating back the power of time over everyday life. This battle would involve terror and death on a massive scale—but from Stalin's perspective, the fate of revolutionary Leninism itself was at stake.

Just as Lenin's attacks on Economism and terrorism would have been pointless without his simultaneous advocacy of an institutional alternative to a politics subservient to spontaneity, Stalin's attacks on Trotsky and Bukharin were combined with his sponsorship of a new type of time-transcending economics: the institutionalization of an economy based on "planned heroism." Overcoming the dichotomy between Lenin's advocacy of draconian labor discipline and his call for enthusiastic time transcendence in economic activity, Stalin hit upon an ingenious synthesis of the two. On a macroeconomic level this synthesis was expressed in the call to "fulfill the Five-Year Plan in four years"; on a microeconomic level it took the form of implementing short-term planning targets and work norms that were supposed to be heroically *overfulfilled* in practice.

"Fulfill the Five-Year Plan in four years!" was the most prominent slogan of Stalin's industrialization drive. In response to this slogan, the various targets of the Five-Year Plan for different industries, which were already based on the extremely ambitious "optimal" targets developed by Vesenkha in 1928, were continually revised upward as the plan progressed. Such practices ensured that the Five-Year Plan would have little operational relevance as a guide to everyday economic decision making at the enterprise level. Indeed, from a Western perspective, Stalinist plan targets made little economic sense at all. As Nove puts it:

> The plan adopted was, to say the least, over-optimistic. Miracles seldom occur in economic life, and in the absence of divine intervention it is hard to imagine how one would expect simultaneous increases of investment and consumption, not to speak of the output of industry, agriculture, and labor productivity, by such tremendous percentages. . . . It is hard to see how anybody could have regarded this as realistic at the time, let alone in retrospect. Yet the optimal variant was shortly afterwards replaced by a still more fantastic series of targets.[38]

In fact, the whole point of Stalin's conception of planning was to make production miracles realistic—and in this way to transcend the limitations of bourgeois economics.

The crucial significance of the Five-Year Plan—as opposed to the yearly and monthly planning targets to which economic activity at the enterprise level was actually geared—was as a symbolic expression on the macroeconomic level of Stalin's conception of disciplined revolutionary time transcendence in economic life.[39] After all, the usual Russian term for "Five-Year Plan," *pyatiletka*, literally means "five-year period." To fulfill a *pyatiletka* in four years thus meant not only to achieve a high growth rate but *actually to compress five years' time into four*.[40] Success on this score would open up the possibility of continual miracles of economic achievement. If five years could be compressed into four, then, in principle, four years could be compressed into three, three into one, and so on. Since this dynamic had no inherent limits, the socialist economy could be seen as possessing practically infinite "reserves" of future growth. The socialist economy would, in theory, thus give rise to an upward spiral of achievement beyond anything witnessed before in human history.

Overfulfilling the Five-Year Plan, from Stalin's perspective, meant literally overcoming the force of time within the time-bound socioeconomic order. And in fact, overcoming ordinary time was precisely what would be necessary to defend the Soviet Union against the capitalist West. As Stalin summed it up in his famous speech to Soviet managers in February 1931:

> It is sometimes asked whether it is not possible to slow down the tempo a bit, to put a check on the movement. No, comrades, it is not possible! The tempo must not be reduced! On the contrary, we must

increase it as much as is within our powers and possibilities. . . . To slacken the tempo would mean falling behind. And those who fall behind get beaten. But we do not want to be beaten. No, we refuse to be beaten! . . .

We are fifty or one hundred years behind the advanced countries. We must make good this distance in ten years. Either we do it, or they crush us.[41]

Given that the entire planning system constructed by Stalin was geared toward the compression of time in economic production—and given the parallel statements about compressing units of time in the works of Marx and Lenin—it becomes clear that the general secretary was not speaking only metaphorically. To "catch up and overtake" the West, the socialist economy would have to harness revolutionary dynamism to bourgeois methods of economic organization. To outstrip time could no longer be considered a theoretical ideal; it had to be made a practical reality.

The Five-Year Plan was thus the charismatic equivalent of Lenin's "revolutionary theory," without which there could be no revolutionary movement. But equally central to the Stalinist socioeconomic system— as to the Leninist political system—was the stress laid on discipline. Clearly, the drive to overfulfill the First Five-Year Plan in four years would be chimerical unless it could be translated into concrete rules governing the daily activity of economic production. The chief mechanisms utilized for this purpose were the instituting of specific work norms used as a standard to judge the revolutionary discipline of Soviet workers and specific monthly and yearly plan targets used as a standard to judge the revolutionary discipline of Soviet managers.

By and large, workers in the Stalinist system were paid according to a system of progressive piece rates.[42] A rather minimal "basic" wage would be paid if a worker merely *achieved* the level of output set by the work norm; above that level piece rates would increase at an accelerated rate as the norm was overfulfilled by a greater amount. In practice, this meant that the work norm assigned to any particular job was no more than an arbitrary marker on what was really a sliding pay scale; a steady achievement of assigned targets guaranteed only a paltry income, even by the impoverished standards of the Soviet 1930s. The key to "making out" in a Stalinist factory was thus to find ways to *overfulfill* one's norms by a substantial amount.[43] However, doing so raised the risk that the norms would be raised in the future, and thus a certain restraint on revolutionary enthusiasm was necessary as well.[44]

The disciplined overfulfillment of work norms was spurred on by the introduction of the campaign for "socialist emulation" during the First Five-Year Plan.[45] Workers taking part in "socialist emulation" strove to outdo the standards of plan overfulfillment set by those at rival enterprises; these enthusiasts were recognized as "shock workers" and held up as models of revolutionary virtue to other, less inspired laborers. In addition, shock workers were called upon to propose "counterplans" that would increase the targets assigned to them by management for a particular shift or a particular planning period; this was supposed to be a crucial element in the drive to fulfill the Five-Year Plan in four years. In return for all this, shock workers were feted in the local press, paid relatively generous wages, and, most important, given the primitive living conditions prevalent in early Soviet factories, provided with special access to special stores, housing, and leisure facilities. In this way, a charismatic-rational form of time discipline—dependent both upon revolutionary enthusiasm and upon a cognizance of the consequences of one's actions in the future—was materially rewarded in the Soviet factory, whereas purely bourgeois time discipline was not.

Finally, the Stalinist system of "planned heroism" was extended to Soviet factory managers as well.[46] Managers and specialists who had become accustomed to the relatively predictable economic environment of the NEP were now confronted with a chaotic drive for continual plan overfulfillment spurred on both from above and below. In addition, managers' salaries, like workers' wages, were based on a sliding scale with special premiums for overfulfillment of plan targets. Predictably, this was an environment to which the old "bourgeois" Soviet managerial class found it difficult to adapt. Indeed, to express what had in 1927 been considered simple economic rationality—the notion that time was inexorable and that present expenditures of labor and capital would inevitably have effects on future production possibilities—became in 1928 tantamount to economic treason. The arrest and trial of fifty-three "bourgeois wreckers" during the Shakhty affair of March 1928, five of whom were executed, sent the clearest possible signal that a fundamentally new sort of outlook on the part of Soviet managers would be demanded along with the drive for increasing production tempos.[47] In essence, factory managers were now expected to become economic heroes with a revolutionary orientation toward time. For this reason, vast numbers of Soviet proletarians were being trained in Soviet schools to fill the shoes of those who had failed to learn to embrace "Russian revolutionary sweep" along with their concern for "American efficiency."[48]

As was the case for Soviet workers, then, the key to success for the Soviet manager working within the Stalinist system was to learn the specifically charismatic-rational form of time discipline. Simple maintenance of time discipline within an enterprise, if it resulted only in plan fulfillment and not in plan overfulfillment, could be taken as a sign that a manager was "hindering the movement for socialist emulation" or making insufficient use of economic "reserves." Conversely, an overly dramatic overfulfillment of planned targets of production would inevitably lead to the raising of the next set of targets, which could then lead to plan underfulfillment in the next planning period. The key to long-term success in this environment, as Berliner has shown, was to learn to overfulfill the plan by only a small amount in each planning period. In other words, successful Soviet managers learned to become socioeconomic "professional revolutionaries" who knew how to transcend time—but in a disciplined way.[49]

This brief sketch of the basic elements of the Stalinist socioeconomic order is not meant to provide a full description of the dynamics of the Stalinist economy. What is crucial to note, however, is the link between the ideal-typical patterns of time use encouraged under Stalin's socioeconomic system and the charismatic-rational conception of time at the core of Marx's and Lenin's previous revolutionary innovations in the theoretical and political cycles. Like Marx and Lenin, Stalin had found a way to oppose both the purely charismatic denial of the need for time discipline *and* the purely rational-legal denial of the possibility of time transcendence in everyday life by holding out an alternative vision of simultaneously time-disciplined and time-transcending action. Unlike Marx or Lenin, Stalin had translated his vision into a set of concrete socioeconomic institutions. And even more remarkably, despite its tragic human costs, its incredible waste and inefficiency, and its unsustainability in the longer run, the Stalinist system of "planned heroism" succeeded in industrializing peasant Russia—the first time in history that an essentially charismatic form of industrial society had been created.

Time, Forward!

It remains to discuss one further important aspect of Stalin's socioeconomic system: the way this system, like Lenin's conception of the party, faithfully reflected the *ethos* of Marx's original revolutionary vision. Just as Lenin had presented the success of his "party of a new type" as a prerequisite for a form of "complete, comradely, mutual confidence among revolutionaries" unattainable under capitalism, Stalin portrayed the collective struggle of the Soviet working class to build socialism, with the party leading the way, as an embryonic form of the universal voluntary community of full communism.

The best illustration of this point is not to be found in the various empirical descriptions of life in Stalinist factories in the 1930s, although such accounts do verify that workers' enthusiasm for socialist construction was a crucial factor in the early days of Stalinist industrialization— even if such enthusiasm often gave way to later disillusionment.[50] Rather, I will attempt to illustrate the essential ethos of Stalinism through an analysis of a novel written during the First Five-Year Plan— the appropriately titled *Time, Forward!* by Valentin Kataev. Though it is based on the author's actual impressions of the industrial city of Magnitogorsk during the early thirties, the real value of this book for the present analysis is precisely that it portrays an idealized version of the Stalinist system.[51] Kataev's novel therefore provides evidence of what party propagandists, following the line laid down by Stalin, were trying to communicate to a mass audience about the essential features of the new socioeconomic order.[52] *Time, Forward!*, in this respect, presents what might be considered the ideal type of the Stalinist orientation toward time in economic production, as synthesized by a pro-Stalin writer of fiction.[53]

Kataev's novel is centered on the struggle of a brigade in Magnitogorsk to break the new world's record for the mixing and pouring of cement recently set by a team of shock workers in Kharkov. The central conflict here is between the hero Margulies, the engineer in charge of the cement brigade's sector, and the villain Nalbandov, the deputy chief of the entire construction. From the outset, Margulies is presented as an extraordinarily time-disciplined man—so time-disciplined, in fact, that he has no need of any external timepiece. Indeed, Kataev describes Margulies excitedly as "an engineer without a watch!":

> Somehow, he could not keep a watch. He would buy one now and then, and he would always either lose it or break it. Finally, he had become accustomed to getting along without a watch of his own. He did not miss it.
>
> He could count time by a multitude of the most minute indications scattered around him in this gigantic, mobile world of the new construction.
>
> For him, time was not an abstract concept. Time was the number of turns of the drum and of the driving pulley; the lifting of the scoop;

the end of one shift and the beginning of another; the firmness of the concrete; the whistle of the machine; the opening door of the dining room; the knitted brow of the timekeeper; the shadow of the plant, passing from west to east and already reaching the railroad track.[54]

Margulies, in sum, is an engineer whose deep concern with time effectiveness has nothing whatsoever to do with the notions of abstract time discipline characteristic of the "bourgeois" West. Rather, he has internalized the concrete time of actual events in order to control the very pace of these events.

In direct contrast to Margulies is the story's villain, Nalbandov. Nalbandov is described as "a brilliant engineer and executive of academic tradition" whose "quickness and decisiveness were grounded in the precise and faultless knowledge of laws whose immutability and sacredness he never doubted himself and permitted no one else to doubt."[55] Nalbandov's fatal flaw is thus not incompetence but rather competence of the wrong sort—an inability to believe in the possibilities of overcoming the "immutable" laws of bourgeois science. His response to the news of the world's record in concrete pouring set in Kharkov is evidence of this: "As far as I'm concerned, I can tell you only one thing, and I say it with absolute finality: I regard all of these experiments as utter foolishness and technical ignorance. This is a construction, not a stunt."[56]

When Margulies at last decides to organize the second shift of the day in an attempt to break the Kharkov record, the stage is set for a dramatic confrontation with Nalbandov. As Margulies's brigade is outside pouring concrete at a breakneck pace, Nalbandov calls the heroic engineer into his office for a dressing down. The ensuing argument between the two is revealing. Nalbandov levels three major accusations at Margulies: that he is violating textbook standards of cement-pouring technique, that he is ruining imported machinery, and that he is wasting limited resources on a futile chase after records. In each case, Margulies's response is that such issues are unimportant in an economy that will overcome the force of time in economic life.

Nalbandov begins with the issue of textbooks:

"I hope that it is not necessary for me to remind you," said Nalbandov without raising his voice, and altogether too calmly, "that less than two minutes for each mixture is against the rules. That is a-b-c. You may find it in any textbook. . . ."

"For us, information contained in any textbook is not compulsory. Every year, textbooks come out in corrected and supplemented editions."

Margulies said it quietly, lispingly, almost in a whisper. . . .

"That's all very well. But during the *current* year, it is recommended that you be guided by the textbooks of the *current* year. Is that not so?"

"Why shouldn't we take advantage of next year's corrections, if we happen to discover them now?"

"Oh, so you want to get ahead of time?"

"We want to finish carrying out the industrial-financial plan."

"It is not timely."

"To go ahead is always timely!"[57]

Nalbandov's second accusation against Margulies is that his "stunt" with concrete pouring is wearing out expensive foreign machinery:

"I say, that with such exploitation the machine is amortised too rapidly."

"In five or six years."

"But, according to the official instructions, it should work, under normal conditions, from ten to twelve years. You are doing violence to a machine."

"Whether it is five or ten years does not matter. Under normal conditions, as you express it, the kind of combine we are building should take eight years to build, and yet, as you are perfectly well aware, we shall build it in three years."

"You may keep your demagogy to yourself. I am calling your attention to the fact that you are contributing to the exceedingly rapid amortisation of imported machinery which has been paid for in *valuta*, and that we do not happen to have dollars scattered on the ground."

"By the time this machine is amortised, we shall no longer have any need of dollars."[58]

Finally, Nalbandov accuses Margulies, who has sped up the concrete-pouring process by ordering the construction of a special wooden planking adjacent to the mixing machine, of wasting precious materials that could be used more profitably in other ways:

"I must tell you that I regard it as utterly uncalled for to waste such a mad quantity of lumber, of which we have a shortage, on experiments of such dubious nature. Why don't you make a hardwood floor? That would make it even easier for your enthusiasts to work. Why don't you put a piano there? Then it will be just like a dancing class."

"If music can lighten our work," said Margulies calmly, "and help us to carry out the industrial-financial plan on time, we shall place a piano there."[59]

What is remarkable about these exchanges is that in each case, it is the "villain" Nalbandov who articulates what must seem plain common sense to anyone with the slightest knowledge of bourgeois economics. Margulies's attempt to break the world's record for concrete pouring clearly *would* violate textbook standards governing the time necessary for mixing concrete in order to maintain its durability; it *would* wear out expensive machinery ahead of schedule; and it *would* waste valuable resources on the efforts of a single work brigade. The novel's presentation of Margulies as a hero thus depends upon the reader's acceptance that "socialist emulation" of the type described would in fact lead to the fulfillment and overfulfillment of the plan, which would in turn create an economy beyond the necessity for bourgeois textbooks, schedules of amortization, and conservation of resources. Margulies's success depends, in other words, on the correctness of something like the following "theory of tempos" articulated by another sympathetic character in the book:

> [The theory] consisted of this: increase of the productivity of one machine automatically entails the increase of the productivity of others indirectly connected with it. And since all machines in the Soviet Union are connected with each other to a greater or lesser degree, and together represent a complex interlocking system, the raising of tempos at any given point in this system inevitably carries with it the unavoidable—however minute—raising of tempos of the entire system as a whole, thus, to a certain extent, bringing the time of socialism closer.[60]

If this "theory of tempos" turned out to be incorrect, however, then the drive to overfulfill production norms in any particular sector of the Soviet economy would have precisely the sorts of negative effects described by Nalbandov.

Such problems, however, do not arise for Margulies because he is not only a revolutionary bent on smashing bourgeois temporal norms. He is simultaneously a disciplined, analytic engineer who knows when to pull back from unwinnable battles against time. In the novel's final chapters, we see Margulies demonstrating the rational component of his revolutionary discipline in his restraining of the enthusiasm of the third shift of shock workers, who arrive on the scene naturally inspired to break the record achieved by their comrades of the second shift. Margulies simply tells them that "it can't be done" without sacrificing necessary quality, that they will be allowed to attempt a new record "in time"—and orders them to work for the time being at slightly less than the record-setting pace just set.[61] Margulies is thus not simply acting on the charismatic strand of Stalinist economics but on the rational strand as well. Scientifically, empirically, he arrives at the conclusion that present conditions do not allow further breakthroughs in the struggle to overcome present conditions.

Such flexibility completely confuses Nalbandov, who, secretly plotting to bring about Margulies's downfall by denouncing him to the plant director, cannot decide whether to accuse the heroic engineer of being a leftist or a rightist:

> It was possible to press two charges against Margulies.
>
> The first: in violation of the exact demands of contemporary science, Margulies *permitted* the number of mixtures to be increased to four hundred and twenty-nine per shift, thereby revealing himself to be a Left extremist.
>
> The second: in violation of the resolution of the Party urging every possible effort to increase tempos, Margulies *forbade* a brigade to make more than four hundred mixtures per shift, thereby undermining the enthusiasm of the brigade and revealing himself as an opportunist of little faith and a Right extremist.
>
> Both charges were just and corroborated by facts. They lacked only motivation: from what motives did Margulies permit on one occasion and forbid on the other?[62]

What Nalbandov cannot understand is that the essence of Margulies's "motivation" is fidelity to the charismatic-rational notion of time—the belief that to be a true revolutionary, one must combine time transcendence and time discipline in one's practical activity.

At the end of the novel, Margulies's approach is, of course, vindicated. But despite Nalbandov's ultimate disgrace in *Time, Forward!*, an important problem remains unaddressed in Kataev's book: can Margulies's method really be sustained in the long run? Can time really be compressed to a greater and greater degree as socialism is constructed, or at some point does the further smashing of "bourgeois" norms of production become, literally, counterproductive?

The tragedy of the socioeconomic system developed in the Soviet Union during the First Five-Year Plan is that on this question Nalbandov, not Margulies, turned out to be right. An economy geared to the overfulfillment of norms and the setting of records by enthusiastic shock workers, coordinated by heroic "engineers without a watch," could not in the long run win the battle against time. *Time, Forward!*, in retrospect, appears to have been remarkably prophetic of later developments in the Soviet economy—precisely because every one of Nalbandov's "bourgeois" complaints turned out, in practice, to be justified. The Soviet economy, from the 1930s until the 1990s, was bedeviled above all by the three problems of the Stalinist model identified by the "villain" of Kataev's novel: poor quality of production owing to the emphasis placed on rush work, the rapid destruction of capital as a result of the Soviet economy's inability to deal with problems of amortization, and the waste of vast quantities of material to facilitate high-priority projects at the expense of a more balanced distribution of enterprise resources.

However, by now it should be clear why Stalin and those who followed him could not simply recognize these problems, as a bourgeois economist might, and deal with them accordingly: to do so would be to admit that an economy freed from bourgeois time constraints was an impossibility—certainly in the backward conditions of Soviet Russia. But this would mean that Lenin's political revolution—given the absence of international proletarian revolution—had itself been a mistake, a premature attempt to break with bourgeois time before the productive forces were ripe. And this, in turn, might well call into question the theoretical synthesis of charismatic and rational time at the core of Marxism itself. If the construction of an economy beyond time constraints was in effect impossible, then capitalist time discipline would seemingly have to be acknowledged as the most productive possible form of economic organization, despite its alienation of the creative potential of human labor.

Such a conclusion was clearly unlikely to be accepted by Stalin and his elite in the process of creating the Soviet "socialist" economy. We must recognize, then, that the specific institutional design of that economy, and the incentive structures that were enforced within it, reflect the worldview that the elite inherited from Marx and Lenin. Communism, according to Marx, was not to be understood as a misty, far-off ideal, but as "the real movement which abolishes the present state of things." Lenin had created a political organization of "professional revolutionaries" whose actual abolishing of the political rule of the bourgeoisie made Marx's words flesh. Following in this tradition, the Stalinist system of "planned heroism" provided heroic managers and workers the possibility of abolishing present time constraints in everyday productive activity, and thus—for the crucial segment of the Soviet population who saw it as legitimate—it too represented a realization of communism in practice. As Kataev puts it in describing the effects of enthusiastic and disciplined work on the shock brigade: "Time flew through them. They changed in time, as in a campaign. New recruits became fighters, fighters became heroes, heroes became leaders."[63] A more succinct description of the essence of Stalinism for the generation of workers promoted to positions of authority in the 1930s can hardly be imagined.

The Cultural Dilemmas of Charismatic-Rational Time

By the end of the First Five-Year Plan, the fundamental components of the Soviet socioeconomic system were in place. Agricultural production was decisively subordinated to the central control of the party-state, albeit at a huge toll in lives. The plan had been declared completed in four years and three months—legitimately, at least as measured in terms of gross industrial production.[64] Meanwhile, new revolutionary cadres had begun to be promoted in the place of the old "bourgeois specialists" in Soviet industry.[65] But now Stalin faced a similar dilemma to that confronting Lenin after the Bolshevik victory in the Civil War: since the "third revolution" had been won and yet neither full communism nor world revolution had arrived, what exactly was the party supposed to do next?

This question, posed now in the environment created by the success of the rapid industrialization and collectivization drives, inevitably brought into clear focus the one issue left unresolved by Marxist theory, Leninist politics, and Stalinist socioeconomics alike: the problem of realizing the mass, cultural internalization of charismatic-rational time norms necessary for the final establishment of the "higher stage of communist society" envisioned by Marx. But while one might design political or socioeconomic institutions that would reinforce the combination of revolution and discipline in practical activity, it was another thing altogether to attempt consciously to create a fundamentally disciplined revolutionary culture.

As we have seen, the entire rationality of Stalinist "planned heroism" was predicated on the idea that the speeding up of tempos in any one sector of the Soviet economy would inevitably work to speed up the tempos of all the other sectors. The unspoken assumption behind this idea, however, was a cultural one: that the heroic norm overfulfillment of a few shock workers would eventually become the prototypical form of labor discipline throughout the Soviet population. Indeed, this was precisely what the campaign for "socialist emulation" was designed to encourage.

But what if, contrary to all expectations, the Soviet workforce proved to be, on the whole, resistant to the elite's attempts to make world records in production a new standard for judging daily work? What if, in fact, Soviet workers interpreted the campaign for "socialist emulation" to be nothing more than a new form of capitalist speedup and saw "shock workers" as norm busters destroying the solidarity of the working class against management demands for increased productivity? In that case, the raising of tempos in one area of Soviet industry would be met not with corresponding heroism in other areas but instead with an increased number of violations of work discipline. This, in turn, would inevitably create bottlenecks throughout the Soviet economy as some sectors lagged behind others.

In fact, precisely this sort of process began to take place almost immediately after the introduction of progressive piece rates in Soviet industry during the First Five-Year Plan.[66] Though at first the genuine enthusiasm of a crucial segment of the Soviet proletariat contributed to real increases in labor productivity in this period, soon de facto bargains between harried managers and increasingly exhausted workers began to be struck. The category "shock worker" became increasingly meaningless as factory managers curried favor with skilled laborers—who were in constant demand in the conditions of labor shortage characteristic of (and partially created by) Stalinist production—by promoting them to this status regardless of merit. *Blat*, or personal connections, became more and more important in determining a worker's position in the factory hierarchy; those with good *blat* were given access to the jobs that allowed for the easiest norm overfulfillment, whereas new workers fresh from the countryside were assigned to jobs with unachievable norms.[67] In short, as it became clear that the fulfillment of the Five-Year Plan in four years was not going to bring about the end of time, charismatic rationalism in economic production quickly gave way to a new socioeconomic phenomenon—corrupt neotraditionalism, now no longer simply within the party but in the factory as well.[68]

Thus, having successfully created a charismatic-rational industrial system along the lines of the charismatic-rational Bolshevik Party, Stalin now faced the problem of having to keep the entire Soviet Union in a state of constant, disciplined revolutionary fervor—or preside over the degeneration of the economy and polity on a massive scale. The same tendency that had transformed Marx's "philosophy to change the world" into a counterrevolutionary Kautskyan dogma and under the NEP had threatened to turn Lenin's "party of a new type" into a corrupt orthodox hierarchy now reasserted itself throughout the Stalinist economic system. Every factory in the USSR was now organized as a revolutionary cell for disciplined time transcendence; every Soviet enterprise had to continue to squeeze the constraints of time or gradually lose its organizational integrity. In the absence of any effective system for declaring enterprises bankrupt or laying off inefficient employees, the routinization of charisma within Stalinist production meant that formal declarations of plan overfulfillment increasingly masked an informal accommodation to "time-serving" economic behavior by workers and managers alike.

Worst of all, from Stalin's perspective, even local party officials became complicitous in the evasion of revolutionary discipline. "Family circles" and "local cliques" protecting one another from the demands of the central authorities began to spring up everywhere.[69] The Bolshevik Party, instead of mobilizing the population for the fight against time, was shielding counterrevolutionary slackers who preferred the "quiet life" to the creation of a new world.

But how could this process be arrested? Stalin had always intuitively known that the integrity of Leninism depended upon successful revolutionary advance. Otherwise, the synthesis of the charismatic and rational components of Marx's vision would begin to break down again in practice. But paradoxically, as the battle to construct a disciplined revolutionary economy in the USSR wound down, the raison d'être of the revolutionary movement seemed to disappear. As Stalin put it at the "Congress of Victors" in 1934, there was "nothing to prove and, it seems, no one to fight," since "everyone [saw] that the line of the Party [had] triumphed."[70] Two years later, Stalin would claim that socialism—meaning the basic foundations of a socioeconomic system based on revolutionary discipline—had been built "in the main" in the USSR; he marked the event by unveiling a new Soviet constitution legally guaranteeing Soviet citizens a wide range of formal freedoms. Logically, there was nothing more for the party to do except preside over the inevitable emergence of communist culture within the Leninist and Stalinist institutional matrix. Yet the culture that seemed to be spontaneously emerging in the mid-thirties was not communist but neotraditional.

Thus, Stalin faced a new battle with the forces of time—now expressed in the increasing desire of proletarians and party members alike to relax rather than work for revolutionary transcendence. Stalin responded to this challenge as he had responded in the twenties: he attacked left, right, and center elements in the party—as well as those he considered to be their "allies" in the socioeconomic elite—in order to clear the way for a new revolutionary advance in the sphere of Soviet culture, the Stakhanovite movement.

However, Stalin's "revolutionary advance" of the mid-thirties had an certain air of unreality about it, for the battles fought in these years were essentially those he had already won. Clearly, Trotsky, Zinoviev, and Bukharin had been utterly defeated. Nonetheless, Stalin continued to claim that even after the successful completion of the First Five-Year Plan, "survivals of capitalism in economic life and in the minds of people" continued to breed confusion and thus "help[ed] to revive the survivals of the ideology of the defeated anti-Leninist groups."[71] In a sense, Stalin argued, the left, right, and center were even more dangerous now than before, because the very success of socialist construction had led party members to let down their guard:

There is the danger that certain of our comrades, having become intoxicated with success, will get swelled heads and begin to lull themselves with boastful songs, such as: "It's a walkover," "We can knock anybody into a cocked hat," etc. . . . There is nothing more dangerous than sentiments of this kind, for they disarm the Party and demobilize its ranks. If such sentiments gain the upper hand in our Party we may be faced with the danger of all our successes being wrecked.[72]

Stalin's conclusion from this was that the need for revolutionary discipline against the pernicious and now hidden influence of the defeated opposition continued more strongly than ever: "We must not lull the Party, but sharpen our vigilance; we must not lull it to sleep, but keep it ready for action; not disarm it, but arm it; not demobilize it, but keep it in a state of mobilization for the fulfillment of the Second Five-Year Plan."[73] The murder of Sergei Kirov on December 1, 1934, by an assassin with a bogus party card in his pocket seemed to provide a dramatic confirmation of Stalin's concerns.[74] Within a month, Stalin's investigation into the murder had identified Zinoviev, the central figure of twenties "orthodoxy," as the primary suspect.

A few months later, Stalin launched a new campaign for charismatic-rational time discipline in the Soviet workplace—Stakhanovism, named after Aleksei Stakhanov, a coal miner in the Donbass who in August 1935 hewed fourteen times more coal than his assigned norm.[75] But just as Stalin's "new" political opponents were the same ones he had defeated in the 1920s, the Stakhanovite movement was very similar to the original campaign for "socialist emulation" and "shock work" launched during the First Five-Year Plan. Once again, the struggle to set new records of production became the focus of industrial activity; once again, those engineers who argued that such feats would merely waste resources, men, and machinery while disturbing the overall coordination of production were branded as "bourgeois formalists" and wreckers.

Stalin's calls for "vigilance" in the party and for Stakhanovism in the workplace were designed to combat the evident desire of an increasing number of Soviet workers, managers, and party members after 1934 to let the revolution take care of itself. Both policies, however, met with continuing resistance from below. The party purge of 1935, rather than restoring "professional revolutionary" discipline and central control over cadres as it was intended to, only served to demonstrate that the party had begun to fragment into hundreds of local fiefdoms.[76] Stakhanovism, meanwhile, encountered enormous passive resistance by managers and workers alike. The same tactics that had inspired genuine enthusiasm among certain strata within the working class during the First Five-Year Plan—though apparently they did have some effect in raising labor discipline—now appeared less inspirational, and more manipulative, to many of those participating in Soviet factory work.[77] Stalin's revolutionary battle against the forces of time was becoming an increasingly lonely one. But a certain segment of the Bolshevik elite was willing to follow him even as this battle took a new form—that of violence directed against the party itself.

Time and the Great Terror

Stalin's policies from 1936 to 1938, despite their horrifying consequences, can be seen as a logical response to these cultural dilemmas of the charismatic-rational conception of time. Faced with the pernicious influence of cultural "survivals of capitalism" in the USSR, as expressed in the evident desire of many in the Bolshevik Party and Soviet population to take time off from disciplined time transcendence, and given the logic of the charismatic-rational conception of time, Stalin could respond in one of three ways.

The first possibility would be to ignore the continuing influence of time within the revolutionary regime in the USSR and to defend the status quo built during the First Five-Year Plan while avoiding any further revolutionary assaults.[78] But such an approach had already been partially incorporated into the party's platform at the Seventeenth Party Congress, including the more moderate targets of the Second Five-Year Plan; it was precisely this relaxation of pressure that seemed to be leading to the rise of *mestnichestvo* (localism) and nepotism within the party. A second consistent response to the dilemma would be simply to admit that the original Bolshevik revolution had not been "timely," since the Soviet proletariat was manifestly not "ready" to live in full communism even within the context of "socialist" political and socioeconomic institutions. But having continually accused Trotsky and Zinoviev of undermining the revolution by articulating a position similar to this, Stalin was hardly likely to adopt it himself. Finally, Stalin could attempt to be faithful to Leninism by continuing the disciplined revolutionary struggle, now directed against the "survivals of capitalism" in Soviet culture themselves. This was, in fact, the path Stalin took.

Still, how could a Marxist justify the continuation of the revolutionary struggle even after the successful building of socialism? Stalin's final contribution to Marxist and Leninist theory provided the answer—the notion that as the final victory of socialism nears, the resistance of defeated enemies *increases*:

> We must smash and cast aside the rotten theory that with every advance we make the class struggle here must subside, the more successes we achieve the tamer will the class enemy become.
>
> This is not only a rotten theory but a dangerous one, for it lulls our people, leads them into a trap, and enables the class enemy to recuperate for the struggle against the Soviet government.
>
> On the contrary, the further forward we advance, the greater the successes we achieve, the greater will be the fury of the remnants of the defeated exploiting classes.[79]

This theory of the intensification of class conflict under socialism is the one theoretical innovation that is universally acknowledged to be Stalin's own. Because the idea seems to be manifestly absurd, no one has tried to claim that it was borrowed from the thought of previous Marxists or Leninists. Indeed, from the perspective of rational, linear time, Stalin's formulation makes no sense. If the class struggle always intensifies as the final victory of socialism approaches, then clearly "socialism" can never actually *arrive*. The revolutionary movement must instead proceed through an endless series of battles against more and more elusive and tenacious "enemies."

This disturbing conclusion is nonetheless completely faithful to the conception of communism as Marx formulated it. According to Marx, revolutionary theory always becomes divorced from revolutionary practice *as soon as* revolutionaries make the mistake of seeing communism as a Utopian "state of affairs to be established" rather than as "the real movement which abolishes the present state of things." Thinking about the postrevolutionary order as a "future" expressed in terms of bourgeois time leads directly either to anarchism, which holds that Utopia will come in the immediate future, or to reformism, which holds that Utopia will emerge "in the long run." Those who think this way, Marx argued, will always be doomed to failure, because time will prove to be stronger than their ideals. For timeless communism to exist at all, it must exist now; it must be an actual movement of time transcendence in time. Communism must find its existence by continually destroying that which exists.

Stalin's formula merely extended this reasoning to the period between the construction of socialism and the emergence of a fully communist society—thus resolving the theoretical contradiction between the "lower" and "higher" stages of communism in the works of both Marx and Lenin. Until the final end of historical time, there could be no relaxation in the struggle against historical time. And it was clear that history was not yet over: the Soviet Union was, Stalin argued, still trapped in a "capitalist encirclement"—a circumstance that some party members, "carried away by economic campaigns and by colossal successes on the front of economic construction, simply forgot about."[80] As long as the external capitalist threat remained, so would the pernicious influence of "bourgeois" habits of thought within the Soviet Union. According to Stalin, there was only one way for a true Marxist-Leninist to approach this problem: through class struggle.

But against whom was the struggle to be directed? Stalin's unleashing of mass terror in 1936–38, though it eventually acquired a dynamic of its own that affected every portion of Soviet society, was at first focused on two major groups of victims: members of the various oppositions to Stalin's line in the 1920s, and managers, economists, and party cadres who had at various points resisted the irrationalities of Stalinist "planned heroism" in economics. Stalin's conscious motives for attacking these groups are, of course, impossible to know. But it seems plausible to argue that in both cases, these first targets of the terror were individuals who, in Stalin's mind, had come to symbolize the chains of time—and who therefore had to be "transcended" through disciplined revolutionary action.

The first group, the Old Bolshevik elite, symbolized for Stalin the constraining power of the past. Zinoviev, Bukharin, the exiled Trotsky, and all their various supporters in the twenties had all witnessed the construction of Leninism and Stalinism *in time*. As long as they lived, the revolutionary movement to abolish time would face one insuperable obstacle—it could not abolish its own history. The Stalinist elite's claim to have made a "great break" (*perelom*) with the past in 1917, and then again in 1928, would always be suspect as long as those who had advocated different policies in the past could testify to the contingency of the policies the party had actually chosen. From the perspective of linear time, of course, it appeared that the left, right, and center oppositions had been decisively defeated; but from Stalin's charismatic perspective, they had become frozen as symbols of counterrevolution. Their protestations of fidelity to Stalin and the party after their defeat could only be seen as hypocritical calumny disguising their "bourgeois" essence. By "proving" in the three great show trials of 1936–38 that his old oppositions—left, right, and center—had "always" been traitorous, and by codifying this "truth" in the new "short course" on party history, Stalin asserted the revolution's ability to change the past along with its continual abolishing of the present.[81]

The second group most severely affected by the terror, the Soviet managerial elite, symbolized for Stalin the constraining power of the future. By advocating the lowering of plan targets, by helping workers resist the extreme pressures of the Stakhanovite movement, or simply by failing to achieve the impossible levels of production that Stalin's plans demanded of them, Soviet managers and economists had demonstrated a concern for the future consequences of present-day policies that violated Stalin's vision of time transcendence in economic life. Even those who had been involved in the explicit measurement of time expenditures in Soviet factories in the twenties and thirties seemed to Stalin to be implicitly suffocating the revolution's dynamism; like Margulies in Kataev's novel, the heroic engineer should not be overly concerned with watches. The fate of the "Nalbandovs" in Soviet industry during the Great Purge was the same as that of the Old Bolsheviks— almost total liquidation.[82]

It may well be the case that in pursuing the logic of the charismatic-rational conception of time to this extreme—by applying it to a "revolutionary struggle" against "survivals of capitalism" in Soviet culture, concretized in the form of individuals who became for Stalin symbolic of the constraining force of time—Stalin revealed himself to be a madman.[83] Perhaps the application of *any* abstract theoretical logic in politics, without regard to the human consequences, is mad. But if Stalin's interpretation of Marxism and Leninism in the mid-thirties was insane, it was apparently a form of insanity that seemed as logical to the party cadres who carried it out as it did to Stalin himself. Again, it must always be emphasized that neither Marxism, nor Leninism, nor even socioeconomic Stalinism made the Great Terror inevitable. But understanding the cultural dilemmas posed by the charismatic-rational conception of time after 1934 can at least allow us to see how a policy of disciplined mass murder could, for a crucial segment of the Bolshevik elite, be a legitimate expression of fidelity to the ideals of the revolution. Indeed, the ultimate tragedy of Marx's vision of communism may not be that it was betrayed by Lenin or Stalin. Rather, the real tragedy may lie precisely in the fact that against all the odds, those who followed in Marx's footsteps interpreted his theory so consistently. No theorist deserves such a fate.

The Fragmentation of Socioeconomic Stalinism: The Post-Stalin Succession and the "Era of Stagnation"

Despite the wholesale destruction of the party, the army, and the managerial cadres during the Great Terror, Lenin's political system and Stalin's socioeconomic system somehow survived. Paradoxically, the greatest external threat ever faced by Soviet socialism—Hitler's invasion of the USSR in 1941—ultimately turned out to be the regime's salvation. The defeat of the German invaders in 1945 allowed Stalin and the Communist Party of the Soviet Union (CPSU) to claim a degree of mass legitimacy, at least in Russia, that had previously been accorded to the regime only by a small minority of party cadres, Red managers, and Stakhanovites.

Even more remarkably, the "Great Patriotic War" appeared to vindicate Stalin's socioeconomic system of "planned heroism." Stalin, after all, had claimed in 1931 that fifty to a hundred years of capitalist development had to be collapsed into ten—"or they crush us." Almost exactly ten years later, the Nazi invasion of Soviet territory began. Furthermore, Stalin's socioeconomic system, built as it was upon the principle of collapsing time in a timely manner, proved to be particularly well suited to the kinds of challenges faced by factory managers and workers in wartime emergencies. After World War II, with the USSR emerging as one of the world's two superpowers, its system of political and socio-economic organization being imposed or replicated on a world scale throughout Eastern Europe and China, Stalin appeared to be a true prophet.

But the newfound cultural legitimacy of Leninist and Stalinist institutions in the USSR, although it sustained the regime for an additional four decades past what might have otherwise been theoretically expected, was not based on the emergence of a communist culture of the sort originally envisioned by Marx, Lenin, or Stalin. Rather, whatever cultural support for the Bolshevik system now existed was decisively traditional in character, based on the regime's demonstrated ability to repel a hated foreign invader of the (Russian) fatherland. This meant, of course, that the cultural dilemmas of the charismatic-rational conception of time continued to haunt the regime. Either the emergence of mass culture of Stakhanovites, or an endless revolutionary struggle against "enemies," or, if neither of these, gradual decay into a stagnant defense of orthodoxy—these were still ultimately the only options for the Stalinist system.

During the remaining years of Stalin's life, all three of these trends were evident. "Heroic" labor and plan overfulfillment was still rewarded with promotions, higher salaries, and Red Banners of the Order of Labor—as such young enthusiasts as Mikhail Gorbachev discovered. The hunt for "enemies" and the emphasis on "vigilance" continued as well; Stalin fortunately died before he could launch a full-scale attack on a new group he began to perceive as human symbols of the constraining power of time, Soviet Jews. Finally, in the late years of Stalinism, the system began to slide inexorably into immobility. The campaigns for heroic labor were gradually tempered by a tacit bargain between the population and the elite that neither would rock the boat too much as long as the system produced both guns and butter at a fairly stable rate.[84] The danger of the corruption of the party-state as pressure from above was eased was mitigated in this period by the memory of the events of the 1930s and by the continuing arrest and imprisonment in labor camps of members of the party and intelligentsia throughout the post-war period. As long as Stalin was still alive, no one could afford to become too complacent about failing at least to *appear* heroically disciplined in everyday activity.

But with Stalin's death in 1953, long-suppressed questions about the nature of the relationship between the Communist Party, the Stalinist state, and Soviet society were unavoidably placed back on the agenda. Having narrowly escaped Stalin's plans for a new blood purge, the post-Stalin Politburo could agree on at least one thing: the need to repudiate Stalin's notion of continual terror as a "solution" to the question of how the party could continue to abolish time even after the construction of socialism "in the main." There would be only one more summary execution of a Politburo member in the USSR: the shooting of Stalin's secret police chief Beria in 1953.

However, the repudiation of terror by Stalin's erstwhile lieutenants reopened the question that had faced the party at the Congress of Victors in 1934: just what should the party be doing during the transition from socialism to full communism? As in 1934, it appeared that there was nothing left to prove and no one left to fight; nonetheless, there was still historical time standing between the Soviet Union and its communist destiny. How could Soviet society be mobilized to bridge this gap?

The key question after 1953, in other words, was the question of how to bring about the cultural internalization of charismatic-rational time discipline within the Soviet population as a whole—without a reliance on continual terror. Only if the masses truly began to act voluntarily like Margulies or Stakhanov, after all, could the inherent dynamism of the socialist form of production be revealed. Their failure to do so could no longer be blamed on the machinations of "enemies of the people." How, then, could a revolutionary but disciplined mass culture be brought into being in the USSR—while preserving the institutional fruits of the revolutionary victories of 1917 and 1928–32?

Remarkably, after the death of Stalin, the "founder" of Soviet socioeconomic socialism, the movement once again split into three camps on this question—in essence, the same three camps that had appeared in the declining phases of the theoretical cycle and the political cycle. The right, represented by Malenkov, stood for a slow, evolutionary approach to creating socialist culture; the left, represented by Khrushchev, stood for a revolutionary denial of time constraints governing this process; and the orthodox center, led by Brezhnev, simply attacked left and right "deviations" while defending Marxist theoretical purity, the "leading role" of the Leninist party, and now the Stalinist socioeconomic status quo as well. Once again, as in the two previous periods of the movement's decline in the Second International and during the NEP, the inability of any of these three leaders to discover a synthesis of charismatic and modern time orientations that would work in practice meant a loss of revolutionary momentum and the increasing corruption of Soviet institutions—ultimately threatening the destruction of the movement itself.

However, because of the changed circumstances created by the success of Stalinist industrialization, there were also crucial differences between Malenkov, Khrushchev, and Brezhnev and their counterparts in the two previous cycles. First, not only did each of these men now accept Marx's revolutionary theory and Lenin's revolutionary party as sacred institutions, but they accorded the same venerated status to the institutions of Stalinist socioeconomics created during the First Five-Year Plan. This is not surprising, since the post-Stalin elite was primarily made up of the very people who had responded to Stalin's call to build a new socialist order in 1928 and who had as a result risen to the pinnacle of world power in a little more than two decades. For these cadres, the Stalinist promise that heroic discipline was redeeming and transformative had been fulfilled perhaps beyond their wildest expectations. In this light, we can see Malenkov as a kind of economically Stalinist Bukharin, Khrushchev as a kind of economically Stalinist Trotsky, and Brezhnev as a kind of economically Stalinist Zinoviev. Though the task was now to create a culture, and not merely a socioeconomic structure, based on a synthesis of time discipline and time transcendence, the historical dilemmas of left, right, and center remained much the same as before.

Malenkov's right, or rational-legal, interpretation of socioeconomic Stalinism was encapsulated in the three policy positions that distinguished him from his main rival for power, Khrushchev: his theory that nuclear war would result in the destruction of socialism as well as capitalism, his reemphasis of managerialism, and his advocacy of placing top priority on consumer goods production over heavy industry. Malenkov's position on the unwinnability of nuclear war appeared to undermine the possibility of revolutionary advance on the international front; like Bukharin, Malenkov appeared to consider a relatively stable international environment a necessary precondition for reformism at home. Malenkov also emphasized the authority of enterprise managers—to whom he, in contrast to Khrushchev, referred as the "commanders of production."[85] Although this managerialism was consistent with his political need as prime minister to consolidate the support of the state hierarchy while deemphasizing the role of the party apparatus controlled by the general secretary, ideologically it represented a "technocratic" position reminiscent of that taken by Bukharin on the issue of how to "rationalize" Soviet industrial production during the NEP.[86] Finally, Malenkov's prioritization of consumer goods over the "production of the means of production" revealed his position on how to maintain the basic institutions of Leninist politics and Stalinist socioeconomics without a reliance on terror and campaigns: the system of "planned heroism" would merely be reoriented toward satisfying the needs of an essentially benign Soviet population. Efficiency and progress in the economy would be ensured by providing Soviet workers material incentives to increase work discipline within the existing socioeconomic structure.

Malenkov's sponsorship of the Soviet manager and consumer and deemphasis of the "heroic" role of heavy industry, combined with his implicit rejection of the possibility of further revolutionary advance abroad, thus represented an essentially Bukharinist, evolutionary response to the further development of the Soviet system in the postsocialist stage. Such an analysis is at least in line with that of Khrushchev himself, who shortly after his defeat of Malenkov in the struggle for power in 1953–54 labeled his rival's positions a "belching forth of the views hostile to Leninism, which in their time were preached by Rykov, Bukharin and their ilk."[87] Given that overcoming the continuing constraints of bourgeois time was now held to be impossible in foreign policy because of the nuclear stalemate between the USSR and the West and inadvisable in domestic policy because of the inefficiency of "heroic" construction projects, it was hard to see what remained of the charismatic element of Leninism in Malenkov's platform. Encouraging better and more efficient work by concentrating on the production of consumer goods had much to recommend it as a practical strategy for overcoming the increasing inertia of the Stalinist planning system. But it was hardly inspiring as an ideal for Bolshevik Party cadres; as during the NEP, Malenkovism appeared to hold out only the prospect of a reliance on rational economic incentives that would lead to full communism only in the very long run of time—if ever. Those who still saw the party's mission as one of time-disciplined time transcendence, therefore, threw their support increasingly behind the party's first secretary, Nikita Khrushchev.

If the views of Malenkov in the post-Stalin era appeared to be in line with those of Bukharin in the 1920s, an examination of Khrushchev's ten years in power reveals the same themes encountered in our discussion of the economic strategy of Lev Trotsky. Khrushchev began his tenure as undisputed leader of the CPSU by reversing all three of the major positions articulated by Malenkov: the idea that a nuclear war would destroy socialism was vehemently denied (although it was claimed that only the capitalists would be foolish and bloodthirsty enough to start one); the revolutionary activity of party cadres in spurring on production in both industry and agriculture was reemphasized; and the absolute centrality of heavy industry in the Soviet economy was reasserted. But Khrushchev soon went beyond an attack on Malenkov's rightism to articulate his own vision of how to bring about the transition from socialism to communism—a vision that was in many ways precisely the opposite of Malenkov's. Khrushchevism was centered on an analytically charismatic belief in the possibility of continuous revolutionary advance on the basis of Leninism and socioeconomic Stalinism.

The prerequisite for this revolutionary advance, according to Khrushchev, was the repudiation of Stalin's terror and the restoration of "socialist legality." Khrushchev's "Secret Speech" to the Twentieth Party Congress of the CPSU in 1956 was the most dramatic possible symbolic statement of his desire—shared by almost all of the post-Stalin leadership—to break with Stalin's strategy for continuous cultural revolution through attacking "survivals of capitalism" within the party and state elite. Stalin's theoretical formula for terror as disciplined revolutionary advance in the sphere of culture, "the closer we are to socialism, the more enemies we shall have," was ridiculed. But significantly, Khrushchev did not revise the notion that Stalin had faithfully followed the correct "Leninist" line until 1934. The socioeconomic revolution of the First Five-Year Plan, including the collectivization of agriculture, was acknowledged to be a great and heroic achievement.

Khrushchev apparently believed that by dramatically and semipublicly revealing the horrors of Stalin's rule from the mid-thirties on, he would clear the way for a massive upsurge in popular enthusiasm for labor and participation in Soviet institutions. The major feature of Khrushchevism in economics was a reliance on "campaignism" to spur on heroic work and time-transcending activity. Just as Trotsky's proposal for "labor armies" in the early twenties called on Soviet proletarians to "display tireless energy" on the production "front," Khrushchev set out to mobilize the entire population of the USSR—within the context of the Stalinist planning system—to engage in the "full-scale construction of communism." The principle that the party represented the vanguard of the "dictatorship of the proletariat" was replaced by a proclamation that the CPSU was now a "party of the whole people." Having built socialism, Khrushchev argued, the Soviet Union had at least potentially created a "socialist" culture extending beyond the narrow confines of the party: "Party workers are not the only politically conscious [*ideinye*] people among us. Each Soviet person—worker, farmer, and member of the intelligentsia—is politically conscious."[88] The entire Soviet population, it appeared, was ready for disciplined revolutionary action.

Khrushchev attempted to unleash the economic potential of this newly "politically conscious" population much as Trotsky had—by launching a series of revolutionary campaigns for economic construction. The Virgin Lands program, the meat and milk campaign, the chemicals campaign—all were predicated on the possibility of overcoming present constraints on Soviet production in a very short time. Finally, in 1959, Khrushchev unveiled the most dramatic expression of his charismatic vision of cultural Leninism—the timetable for reaching the stage of "full communism" by 1980. There is no reason to believe that Khrushchev was insincere about this proclamation. In many ways, his position was a resurgence of the leftist tradition within the Marxist movement from the Second International on—a belief that true Marxism required a continuous process of "permanent revolution," with little or no emphasis placed on any "realistic" assessment of present time constraints on this process.

But just as Luxemburg's charismatic Marxism foundered on the unwillingness of empirical proletarians to sacrifice their material security for socialism, and just as Trotsky's charismatic Leninism quickly ran up against the empirical backwardness of the Soviet peasant economy, Khrushchev's hopes of a rapid overtaking of the capitalist West on the way toward full communism in two decades proved to be wildly overoptimistic when judged from a rational, linear time perspective. The Virgin Lands program, after a few years of impressive success, proved to have been based on an inaccurate assessment of the long-term fertility of the newly planted regions; the campaign to overtake the West in milk and meat production led to the wholesale slaughter of livestock in 1959 followed by a huge dropoff in these areas in subsequent years. As any bourgeois economist could have predicted, heroic attempts to overcome time constraints in production in the short term led to even larger losses in the long term.

Like Trotsky, Khrushchev blamed the increasing failure of campaignism as a strategy of economic growth on the stagnant Soviet "bureaucracy"; also like Trotsky, he appeared to base his opposition to bureaucracy on a deeper hostility to institutionalization in general. This, too, reflects a fundamentally charismatic time orientation; Khrushchev hated to see Soviet reality get fixed into frozen, time-bound forms. Thus, as the first secretary's reign proceeded, party and state institutions were continually rearranged, beginning with the introduction of regional economic councils (*sovnarkhozy*) in the place of the central ministries in 1957 and ending with the bifurcation of the party itself into a section overseeing agricultural production and a section overseeing industry in 1962. The combination of Khrushchev's failures in economic policy and his erratic reorganizations of the administrative structure of the USSR led to his dismissal as party leader in 1964; Khrushchev, like Luxemburg and Trotsky before him, spent his final days marginalized from the mainstream of party orthodoxy—a logical result of the attempt to apply a pure form of time transcendence in political practice.

Thus, as in the previous theoretical and political cycles, the policies of Malenkov's right appeared to promise only a bourgeois form of success, and the policies of Khrushchev's left promised only a genuinely revolutionary form of failure. Brezhnev's centrist alternative to these positions from 1964 to 1982, like Zinoviev's platform in 1924, was simply to resist all pressures for change from the left and the right, defending the ideological, political, and now also the socioeconomic status quo. Thus, the "voluntarist" Khrushchev was lambasted as a "hare-brained schemer" whose constant reorganizations of party and state institutions had produced nothing but chaos.[89] Though the early years of Brezhnev's rule saw a brief resurgence of Malenkov-style economic policies as embodied in the pro-manager, pro-consumer "Kosygin reforms" of the mid-sixties, such "right deviations" toward market socialism were curtailed throughout the Soviet bloc after the crushing of the Prague Spring in Czechoslovakia in 1968—an event that seemed to prove the dangers of rightist reformism in Eastern Europe and at home. Official intolerance for every alternative interpretation of Leninism led to the rapid growth of a dissident movement including even many intellectuals favorably inclined toward socialism—a movement that was increasingly severely suppressed during the 1970s.

From 1971 on, Brezhnev's rule became marked by the elite's increasing reluctance to tolerate even the slightest deviation from the theoretical and institutional legacy of Marx, Lenin, and Stalin. The Khrushchevian notion of the "party of the whole people" had quietly been dropped in favor of a stress on the CPSU's "leading role" under conditions of "developed socialism"—a formula whose emphasis on the past tense perfectly summed up the ethos of Brezhnevian orthodoxy. The Leninist party and the Stalinist economy of "planned heroism" were seen as the concrete manifestations of revolutionary time discipline; any further attempts at time transcendence in action, as Khrushchev's "scheming" had demonstrated, could only undermine the stability of the "revolutionary" order. Economic reform, to the extent it was introduced at all, was now designed merely to "perfect" the Stalinist planning system.[90] Even Stalin himself began to be restored to a place of honor in the Leninist pantheon; the most tangible manifestation of this was the late dictator's reburial in a conspicuous place behind Lenin's tomb in Red Square.

As had been the case with Kautsky and Zinoviev before him, Brezhnev's validation of this centrist interpretation of charismatic rationalism rested on his portrayal of the USSR as an international fortress and orthodox center for the world revolutionary movement. This validation was explicitly articulated through Brezhnev's standard opening speech to party congresses—a recitation of the list of Third World countries that had, in the last five years, either joined the "socialist camp" or adopted a "socialist orientation."

Unlike Zinoviev, whose attempt to defend Leninist orthodoxy foundered because of his lack of political skill and his absence of vision on socioeconomic issues, Brezhnev instituted a form of neotraditional rule that was embraced by significant segments in the party and society. By 1964, the Soviet Union had experienced a political revolution, a civil war, a socioeconomic revolution, a cultural revolution against "survivals of capitalism" in form of the Great Terror, a devastating world war, and a decade of "hare-brained" institutional reorganization. Throughout this period, the members of the Brezhnev Politburo had responded to the Bolshevik Party's call for heroic advance. To the Leninist "heroes" who had created Soviet socialism, beaten the Nazis, and achieved parity with the capitalist superpower, the United States, the idea that it was time simply to preserve what had been achieved in the century-long battle to master time—including their own positions of power—was an appealing one.

As in the period of the Second International, as during the NEP, and as during the Second Five-Year Plan, this relaxation of revolutionary time discipline led to a gradual degeneration of charismatic rationalism into corrupt orthodoxy. Personal "connections" (*blat*) once again were tacitly allowed to play the dominant role in the state economy; the battle to overfulfill planning targets was replaced by widespread falsification of reporting of the levels of production achieved by Soviet enterprises; and the Stakhanovite ideal for Soviet labor seemed increasingly empty in a society based on the principle of the well-known Soviet joke: "We pretend to work, and they pretend to pay us."

However, the neotraditional form of Leninism established in the Soviet 1960s and 1970s proved to be much more stable than the defense of orthodoxy in the previous cycles of Marxism's and Leninism's development. The fact that Stalin had constructed a functioning, if erratic, charismatic-rational industrial economy in the USSR meant that the decay of revolutionary momentum under Brezhnev could continue for two decades without threatening the survival of the Soviet state. By the 1980s, however, the Soviet Union had become mired in an "era of stagnation" from which it would never really recover—despite the efforts of Mikhail Gorbachev to reclaim Marx's and Lenin's legacy by launching the disciplined revolutionary campaign to produce socialist culture known as *perestroika*.

6

GORBACHEV'S PERESTROIKA AND THE CHARISMATIC-RATIONAL CONCEPTION OF TIME

ikhail Gorbachev, in his tenure as general secretary of the CPSU from 1985 to 1991, irreversibly changed the course of Soviet history. However, the problem of how to conceptualize the precise nature of this change has proven to be a thorny one. The two paradigms that dominated Western analysis of the Soviet Union—the totalitarian model and modernization theory—were each faced with enormous difficulties in making sense of the Gorbachev *perestroika* reforms. The totalitarian model, which in its original form insisted that the Soviet system was inherently immutable and driven toward absolute power, simply proved incapable of accommodating the mounting evidence that Gorbachev was not a "refined Stalinist" but a man committed to changing the very nature of party rule in the USSR.[1] The destruction of the Stalinist system, which the totalitarian model insisted could occur only if power was wrenched from it, had contrary to all expectations been initiated by the party elite itself. Although the collapse of communism in 1989–91 had in a sense vindicated the totalitarian assumption that a "reformed Leninism" was essentially an impossibility, the model had failed to provide any satisfactory explanation why the Gorbachev Politburo itself had led the headlong rush toward Leninism's worldwide dismantling.

Modernization theory, on the other hand, readily explained why Gorbachev had initiated "radical reform" in the USSR: he was responding to the political "imperative" in an industrialized society to introduce norms of political democracy and market efficiency.[2] Gorbachev's ascendancy could be seen as the final victory of the "reformers" in the Soviet elite in their long battle with ideological "conservatives" and the long-awaited emergence of a fully "modern" leader in a "deideologized" CPSU.[3] Gorbachev's constant use of Leninist phraseology in his advocacy of *perestroika* could be explained as a necessary concession to the old norms of party rule—a concession that would become unnecessary once the power of the apparat had been broken. But modernization theory, having identified Gorbachev's *perestroika* with the cause of modernity in the Soviet Union, was subsequently unable to make sense of the increasing marginalization of *perestroika* as a political platform and, by 1991, the disintegration of the Soviet Union itself.[4] If Gorbachev represented the triumph of modernity and rationality in Soviet politics, why did he at the last moment recoil from the implications of the "500-days plan" for full marketization of the Soviet economy advocated by Boris Yeltsin and some of his own economic advisers? Why, even after the August coup of 1991, was he still insisting on preserving the essential elements of the "socialist choice" in a reformed USSR? And if Soviet society had become essentially "modern" through industrialization, urbanization, and education, why did the collapse of Leninism seem to be leading to prolonged social turbulence and political fragmentation, rather than allowing for a peaceful evolution toward Western norms of liberalism and capitalism?[5]

Both paradigms, then, explained at best only part of the Gorbachev story. Modernization theory had predicted the eventual emergence of a Leninist reformer who would attack the Stalinist system, and thus it was plausible as an explanation for Gorbachev's initial reformism. But it had trouble explaining why that reformer would cling to his outmoded ideology long after it was politically expedient to do so. Totalitarian theory had argued that no reform remaining within the ideological confines of Leninism could succeed, and thus the ultimate collapse of *perestroika* made sense. But it had trouble explaining how *perestroika* could have been launched by the party elite in the first place, given Gorbachev's apparent continuing ideological commitment to Marxism-Leninism.

As had been the case with scholarly analysis of the historical link between Leninism and Stalinism, and of that between Stalinism and the post-Stalin era, analysis of the Gorbachev period tended either to insist on the essential immutability of the communist system or to overemphasize the elements of change in Soviet society, in both cases presenting a one-sided picture of the Gorbachev phenomenon. Gorbachev, who was simultaneously a committed Leninist *and* the major force behind policies that destroyed the existing system of party rule in the USSR, was inexplicable from either point of view.

Given the theoretical framework and historical overview provided here, we are in a position to approach the analysis of the Gorbachev era from a perspective somewhat different from that of the totalitarian model or modernization theory. As with the question of Lenin's connection to Marxism and Stalin's connection to Leninism, we must neither deny the thread of historical continuity linking Gorbachev to the Soviet past nor exaggerate the "inevitability" of a "radical reformer" taking over the reins of the party-state. A Weberian analysis of the connection between Gorbachev and the entire past history of the Marxist revolutionary movement must, on the one hand, demonstrate the elements of continuity between Lenin, Stalin, and Gorbachev that can be shown to exist in ideal-typical terms, but it must, on the other hand, remain true to the contingencies of historical detail.

I argue in this chapter that Gorbachev must be understood not as a "refined Stalinist," nor as a closet "modern" who disguised himself long enough to rise to the position of leader of the CPSU, but as a would-be innovator within the context of the charismatic-rational conception of time in the mold of Marx, Lenin, and Stalin. Specifically, Gorbachev's *perestroika* was introduced as an attempt to bring about the rapid transformation of Soviet culture in a charismatic direction—to produce a culture, and not merely a socioeconomic structure, based on a mass internalization of norms of disciplined time transcendence in everyday life. I analyze the three major stages of Gorbachev's efforts to move the Soviet Union in this direction—the early period of "acceleration" (*uskorenie*), the middle period of "restructuring" (*perestroika*), and the final period of the rejection of Leninist norms of party rule—as a logical policy sequence given Gorbachev's analysis of the nature of Soviet society and his professed desire to reinfuse the institutions of Soviet socialism with new revolutionary content. Finally, I argue that unlike the innovations of Marx, Lenin, and Stalin, Gorbachev's mechanisms for achieving a charismatic-rational attitude toward time in Soviet mass culture were doomed to failure—a failure that meant the final destruction not only of Gorbachev's vision but of the economics of Stalinism and the politics of Leninism as well.

Gorbachev's Early Career

Before examining the three stages of Gorbachev's reform strategy, we must first address the question: How was it possible that a "true believer" in the Marxist ideal of charismatic rationalism in time orientation, one who aimed to remake Soviet culture along charismatic lines, could emerge as leader of the CPSU after two decades of stagnation and corruption of the party-state under Brezhnev? To answer this question, it is necessary to emphasize the particular set of life experiences that uniquely shaped Gorbachev and his generation within the party.[6] Here, three factors are crucial to an explanation of the idealistic mind-set of Gorbachev and his advisers: their teenage experience of Soviet victory in World War II and postwar reconstruction, an early adulthood in the heady days of Khrushchevian optimism about the Soviet future, and long and bitter years of marginalization from the center of party life under Brezhnev during the years of political maturity.

Stalin's victory in World War II, the emergence of the Soviet Union as a superpower, and the establishment of Leninist regimes in Eastern Europe were the key factors in the legitimation of Stalin's socioeconomic charismatic rationalism on a mass level in Soviet society. Certainly, no generation felt this more keenly than Gorbachev's. Fourteen years old in 1945, Gorbachev came of age in an environment marked by heroic mass struggle and sacrifice under Leninist auspices. Unlike the generation of Khrushchev and Brezhnev, whose political outlook was shaped by the struggle to establish the socioeconomic institutions of Stalinism in a hostile social environment and who therefore saw victory in the war mainly as a decisive legitimation of the Stalinist *socioeconomic* system, Gorbachev experienced the war years and the postwar reconstruction quite differently—as a realization of the Leninist *cultural* ideal of complete heroic participation of the masses together with the party in a common cause.[7] The older generation of leaders had been promoted not only for their acceptance of norms of revolutionary time discipline in Soviet institutions but also for their ruthlessness in fighting kulaks, bourgeois wreckers, and enemies of the people; they were not likely to be persuaded that the cultural ideals of communism could be achieved in the absence of the central leadership of the party-state. Gorbachev, on the other hand, first experienced Leninism in a context of wartime unity between party and people, which made the vision of a rapid diffusion of the norms of disciplined time transcendence throughout Soviet culture seem not a vain dream but part of historical experience.[8]

The idealism of Gorbachev and his generation was solidified by the experience of party work under Khrushchev. Khrushchev's own charismatic solution to the problem of creating a cultural basis for communism in the USSR closely paralleled the vision of Leninism characteristic of the eager and idealistic Gorbachev, now in his mid-twenties. Of course, Khrushchev himself never questioned the fundamentals of the Stalinist planning system; he was convinced that inspiring campaigns for economic achievement, and not Stalinist coercion, would allow cultural charismatic rationalism to take root in the USSR within the context of Stalinist planning and norm setting. But in his call to establish "full communism" by 1980, Khrushchev validated on a political level the vision of unity between party and society that Gorbachev had experienced during and immediately following World War II.

Khrushchev's ouster by the Brezhnev oligarchy in 1964 could only have been seen, by Gorbachev and those like him, as a betrayal of a realizable and noble vision. Unlike the older generation of Brezhnev, Gorbachev had not personally witnessed the failure of similar promises of the rapid victory of communism in the early years of the revolution; he had no internalized appreciation of just how hard fought the battle had been for the establishment of Stalinist economic institutions in the early 1930s. That Khrushchev's rule threatened the continuation of the established Soviet economic system therefore must have seemed less important to Gorbachev than it did to the older generation; what resonated for the Gorbachev cohort within the party was Khrushchev's vision of the achievability of communism—a vision on which Brezhnev had seemingly turned his back.

Such an impression could only have been strengthened by the years of party decay and loss of momentum under Brezhnev's leadership, which Gorbachev, like most of his future advisers, watched from the periphery of the Soviet Union, in places like Stavropol, Sverdlovsk, Tomsk, and the Transcaucasus. With no opportunity to test their own ideas about furthering the revolutionary cause of Leninism, these men were left with an absolute hatred of ossified central control and the idealistic belief that its destruction would by itself lead to a rapid realization of the cultural ideals promulgated by Khrushchev.

Brezhnev's neotraditional rule from 1964 to 1982 therefore had a different effect on the Gorbachev generation than on the older generation of party members or on the younger generation born after Stalin's death. Brezhnev's stress on party and state orthodoxy led among the older generation to increasing corruption and a loss of interest in further revolutionary advance, as they began to identify the maintenance of their own power with the defense of Leninism itself.[9] The younger generation born after World War II, who had seen nothing but the period of Leninism's decline, were not so much corrupted—since they had never internalized the values of revolutionary Leninism in the first place—as they were alienated from the regime over time. But for the small group who shared the political and life experiences of Gorbachev, Brezhnev's rule absolutized a particular type of Leninist idealism—an idealism based on a deeply held belief that mass, disciplined time transcendence was an achievable cultural goal and that Soviet society, like the Gorbachev generation itself, awaited only the destruction of Brezhnevism for its realization.

The Gorbachev cohort was thus, by the 1980s, the last substantial group in Soviet society both to believe in the original ideals of Marxism and Leninism *and* to remain substantially uncorrupted.[10] The idealism characteristic of party members of Gorbachev's generation had prevented them from using their party positions merely as tools for furthering personal power and privilege and led them to despise the large number of cadres who did. This is what ultimately created a political opening for Gorbachev and those who shared his outlook. To understand how this previously marginalized group of idealistic Leninists rose to positions of central power in the USSR, we must briefly examine the history of the immediate pre-Gorbachev period, the Andropov-Chernenko interregnum.

Yuri Andropov's background was quite different from Gorbachev's. Born in 1914, he had been old enough to appreciate the struggles of Stalinist industrialization and collectivization and showed no inclination during his brief reign as general secretary to attack the fundamental institutions of the Soviet economy. But his position as KGB chief during the worst years of Brezhnevism's corruption gave him a unique appreciation of just how severe the situation within the party, and how frequent the "violations of labor discipline" by Soviet workers, had become by 1982. Determined to reverse these trends, Andropov began to attack party corruption and worker inefficiency alike under the auspices of an all-out "campaign for discipline."

How would Andropov's policies have progressed if he had lived? The idea of a mass discipline campaign to improve Soviet economic production was not new; it was, in fact, merely a reiteration of faith in the Stalinist charismatic-rational conception of time in socioeconomic affairs, stressing the need to combine campaign-style enthusiasm with disciplined work. But Andropov provided no clear response to the cultural dilemmas that had made the Stalinist synthesis of discipline and enthusiasm increasingly unworkable in the post-Stalin era. The early successes of the antialcohol and anticorruption campaigns were bound to fade as the most egregious violations of party and work norms were eliminated; Andropovism would sooner or later have run up against the same difficulties that Gorbachev was to encounter in his second year in office, as cultural opposition to the further tightening of discipline and the stricter enforcement of sobriety began to manifest itself. Whether Andropov would have moved in a direction similar to Gorbachev's, slid back into Brezhnevian tolerance, or increased the use of secret police coercion in party and economic affairs—thus raising the specter of a return to Stalinism—is hard to say.

What is crucial to emphasize is the immediate political effect of Andropov's brief period of leadership: the promotion of the Gorbachev generation of "Leninist Romantics" to top party posts.[11] Striving to build a power base within the party that would be untainted by corruption, Andropov found allies in the one group who remained committed to the possibility of a cultural realization of the ideals of Marxism-Leninism. Andropov counted on the hatred of corruption, and faith in the revolution, of these younger party members. What he could not have foreseen was that the discipline and faith of Gorbachev and his cohort would later fuel a fanatical effort to reform Leninism along the lines of these cultural ideals—even at the cost of party rule itself.

If Andropov's tenure as general secretary led to the promotion of Gorbachev and members of his generation to key party positions, Konstantin Chernenko's even briefer reign put the finishing touches on the political discrediting of Brezhnevism. While Brezhnev could at least claim to be preserving political and socioeconomic orthodoxy while awaiting the revolution's ultimate international victory, Chernenko's *return* to Brezhnev's policies after Andropov's promised purification of the movement seemed a final repudiation of the party's claim to be revolutionary in its essence. The same old guard that had maintained a somewhat unexamined support for Brezhnevism as representing the triumph of "real socialism" were now forced to deny any such legitimacy to the ailing Chernenko's embarrassing (and again, embarrassingly brief) period of leadership. Looking within the existing party elite for a new source of inspiration and revolutionary rejuvenation, the CPSU leadership, still dominated by Brezhnev appointees, settled on one of Andropov's youngest and most prominent protégés, Mikhail Gorbachev.

Humiliated by two decades of increasing corruption, declining Soviet international prestige, and finally the spectacle of three general secretaries dying within two and a half years, the old guard allowed political changes in Gorbachev's first few years in office that would formerly have been unacceptable to them. Within a few years, every member of the last Brezhnev Politburo had been ousted from the leadership, and Gorbachev had set out to remake the political and socioeconomic systems of the USSR according to an absolutized ideal of cultural charismatic rationalism in time orientation.

From *Uskorenie* to *Perestroika*: Gorbachev's First Two Years

Gorbachev was named general secretary of the CPSU on March 11, 1985. From that point until the Twenty-seventh Party Congress of the CPSU in February 1986, he was mainly engaged in the struggle to consolidate his political authority, purging holdovers from the Brezhnev era in the party apparatus and promoting his own supporters. By the time the Twenty-seventh Party Congress met, nearly 40 percent of the Central Committee was newly elected.[12] In economic policy, Gorbachev pushed for a new attack on alcoholism among Soviet workers and for a strengthened enforcement of labor discipline. In short, the new general secretary made it clear that he was following his mentor Andropov in attacking the corrupt, neotraditional form of Leninism characteristic of the Brezhnev period. But there was little indication that Gorbachev would go beyond these policies and begin to attack the institutional structure of Leninism itself.

By the time of the party congress, however, Gorbachev felt secure enough in his position to begin articulating a distinctive "Gorbachevian" interpretation of the current tasks of the party—an interpretation that broke more decisively with Brezhnevian Marxism-Leninism than anything put forth by Andropov. This new interpretation of party ideology had three major components: a relativization of the international arena as a realm for the legitimation of Soviet rule and a corresponding reemphasis on domestic politics, an insistence that the Soviet Union was at a "turning point" (*perelom*) in its development, and a call for the "acceleration" (*uskorenie*) of socioeconomic development as the proper response to this challenge.

As we have seen, the downgrading of the importance of international socialism as a sphere for the legitimation of party rule had been an important aspect of the previous innovations within the framework of charismatic rationalism in time orientation by Lenin and Stalin. Lenin, in his *Imperialism*, had broken with Kautsky's centrist internationalism, which was based on the preservation of Germany's superior standing within the Second International as a concretization of revolutionary orthodoxy, by establishing the rival doctrine that a proletarian revolution in backward Russia would break the structure of imperialism at its weakest point. Likewise, in 1924 Stalin had broken with Zinoviev's similar form of internationalism, during the latter's period of leadership of the Comintern, with his development of the doctrine of "socialism in one country."

Gorbachev's break with Brezhnev's neotraditionalism closely paralleled these two earlier leaders' attempts to establish themselves as Marxist innovators on the Russian scene rather than orthodox centrists focusing their attention on international affairs. Thus Gorbachev's report to the Twenty-seventh Party Congress broke with the long-standing practice under Brezhnev of starting out each congress address with a chronicle of recent socialist victories in the Third World. Gorbachev began his speech with an unprecedented portrayal of the "complex" state of world affairs, printed in official documents under the subheading "Tendencies and Contradictions." Despite Gorbachev's stern denunciations of the "imperialist system . . . still living off the plunder of the developing countries, off their totally merciless exploitation," the major thrust of his analysis was the need for "constructive and creative interaction between states and peoples of the entire world."[13] The section of Gorbachev's speech dealing explicitly with international affairs—relegated to near the end of the general secretary's report—made it clear that he saw little hope of further socialist advance abroad without prior reform in the Soviet Union: "Today the destinies of peace and social progress are tied up more closely than ever with the dynamism of the economic and political development of the socialist world system. . . . Both friends and enemies look upon us, the immense, many-faced world of the developing countries looks upon us, seeking its choice, its path. What this choice will be depends to a large extent on the successes of socialism, on the credibility of its responses to the challenge of the times."[14] The message was clear: in an exceedingly complex world, the progress of socialism on a global scale depended on the Soviet Union's response to its own internal problems.

If Gorbachev's reemphasis of Soviet domestic politics over international affairs prepared the ground for his advocacy of change, his argument that the Soviet Union had reached a "turning point" (*perelom*) in its development, one analogous to earlier periods of revolutionary advance in Soviet history, showed his desire to usher in a new cycle of Leninist development. Just as Lenin had broken with Luxemburg's left, Bernstein's right, and the Kautskyan center in establishing a Marxist political regime, and just as Stalin had broken with Trotsky, Bukharin, and Zinoviev in establishing a Leninist socioeconomic regime, Gorbachev now explicitly broke with the Brezhnevian status quo:

> For a number of years the deeds and actions of party and government bodies trailed behind the needs of the times and of life. . . . The problems in the country's development built up more rapidly than they were being solved. The inertia and stiffness of the forms and methods of administration, the decline of dynamism in our work, and an escalation of bureaucracy—all this was doing no small damage. Signs of stagnation had begun to surface in the life of society.
>
> The situation called for change, but a peculiar psychology—how to improve things without changing anything—took the upper hand. . . . But that cannot be done, comrades. Stop for an instant, as they say, and you will fall behind a mile.[15]

Gorbachev's insistence that 1986 marked a *perelom* in Soviet history therefore presented the party a stark choice: undertake "radical reform" of Soviet institutions, or watch the party's last chance for revolutionary renewal slip by.

In essence, Gorbachev's analysis paralleled Lenin's battle cry to put an end to the stagnation of the Second International by undertaking revolutionary action, as well as Stalin's famous call to "catch up and overtake" the capitalist countries in ten years or be crushed by the international bourgeoisie. The difference was that with political and socioeconomic institutions already controlled by the Leninist party-state and organized along charismatic-rational lines, the only fortress left for the Bolsheviks to storm was the fortress of linear time's dominion over Soviet culture itself. Thus Gorbachev, from the outset, operated with an ideal of success for his reform effort that was—from the Weberian perspective—unrealizable: the realization of a mass cultural norm of permanent, disciplined revolutionary time transcendence in Soviet society.

This vision was encapsulated in the early Gorbachev slogan *uskorenie*, or acceleration. As a criterion for economic progress, *uskorenie* differed from the typical liberal capitalist standard of success in crucial respects. Whereas elites in capitalist regimes tend to judge themselves by whether they can sustain a simple growth rate in gross national product, assuring themselves that this means they are making "progress" in linear time, Gorbachev was calling for the creation of an economic mechanism which would, in principle, result in *continually increasing growth rates*—2 percent one year, 4 percent the next, 6 percent the next, and so on. If simple growth allowed capitalists to feel they were keeping pace with time, "acceleration" would guarantee to Leninism an eternal dynamism in Soviet life that would ultimately usher in a qualitatively different sort of social order. It was in this sense that Gorbachev could argue that while acceleration meant "first and foremost, raising the economic growth rate," it did not "amount only to a transformation in the economic field" but actually held the key to all the Soviet Union's problems "in the near and more distant future—economic and social, political and ideological, and internal and external ones."[16]

Very quickly, of course, such a standard of success would work to undermine Gorbachev's authority, as the promised renewal of socialism failed to take place. But it is crucial to realize that the charismatic component in Gorbachev's time orientation—his sense that a rapid overcoming of current time constraints was possible through correct revolutionary action—was from the outset at least equal to his analytically rational-legal emphasis on scientific analysis of the complexities of the Soviet Union's crisis and the need for a disciplined approach to overcoming it.

In short, Gorbachev's time orientation, as displayed in the address to the Twenty-seventh Party Congress, was charismatic-rational in the classic Marxist-Leninist sense. The new general secretary, in focusing on Soviet domestic politics as the proper arena for revolutionary endeavor, in breaking with the neotraditional stress on orthodoxy characteristic of Brezhnev, and in outlining an alternative vision of *uskorenie* as the key to unlocking the inherent dynamism of the socialist economy, sounded themes that had real meaning for cadres steeped in Leninist norms of discourse and practical activity. Indeed, Gorbachev's success in convincing a significant percentage of the party establishment to back reform policies that from the beginning threatened their basic material interests can only be explained by understanding how Gorbachev's program resonated on the level of Leninist *ideals*. After years of unrealizable Utopianism followed by decades of unrevolutionary defense of the status quo, Gorbachev's call for a "radical reform" of Soviet institutions at last promised the party a chance to do something simultaneously heroic and realistic.

But just how this promise would be translated into an actual program for institutional change was not clear in 1986. At first, Gorbachev's economic program remained very similar to Andropov's in its stress on increased labor discipline, sobriety, and crackdowns on corruption. As during Andropov's reign, this set of policies produced short-term payoffs as the most egregious violations of norms of work and party activity were attacked. But by the beginning of 1987, there were already signs that these methods of stimulating economic production had begun to bog down. Sugar began to disappear from store shelves as the production of *samogon* (moonshine) skyrocketed in the absence of state-supplied vodka. The arrest on charges of corruption of Brezhnev's son-in-law, Yuri Churbanov, rather than presenting the image of a newly cleansed party apparatus, had the effect of demonstrating that the CPSU as an institution had become rotten to the core. At the same time, efforts to enforce labor discipline and efficiency in time use in the Soviet economy began to run up against inherent limits. Early attempts to introduce a three-shift system in Leningrad factories, in order to keep machinery in operation twenty-four hours a day and thus in principle overcome problems of idleness in industrial production, were abandoned as complaints poured in from workers about the impossibility of living normally on the night shift in the absence of twenty-four-hour grocery stores and nighttime child care.[17] The Andropov strategy for enforcing cultural Leninism—mass disciplined enthusiasm for work *within* Leninist institutions—had failed.

The beginning of 1987 found Gorbachev at a crossroads. He could intensify the campaign for discipline, but with the most obvious violations of discipline already under attack, this would have necessitated a neo-Stalinist reliance on punitive sanctions and potentially large-scale police violence against Soviet workers that directly contradicted Gorbachev's professed anticentralism. At the same time, Gorbachev had decisively rejected both the Khrushchevian strategy for eliciting work enthusiasm from the Soviet population—charismatic promises of a quick end to economic scarcity—and the Brezhnevian attempt to enforce political, social, and economic orthodoxy despite increasing cultural anomie. There remained the "right" option: explicitly to reject the possibility of achieving socialist cultural transformation except perhaps in the very long run and to concentrate on introducing true marketization and on producing or importing more consumer goods to inspire Soviet workers to work harder. But when this strategy was articulated by the prominent economist Nikolai Shmelev in the summer of 1987, it, too, proved to be unacceptable to Gorbachev.[18] Like Bukharin's rightist strategy for economic development in the 1920s, Shmelev's proposals relied ultimately on the specter of unemployment as a tool for enforcing work discipline where positive work incentives had failed, and this was incompatible with Gorbachev's platform. Gorbachev had promised socialist renewal; adopting the right-wing strategy would have meant accepting the necessity of yet another long-term period of retreat, which would no doubt have been ruinous for a party already substantially corrupted and in danger of losing its revolutionary ideals altogether.[19]

The "Human Factor"

Faced with these rather unpalatable choices, Gorbachev began to rely increasingly on a group of close advisers who had worked at the Institute of Economics and Industrial Organization at the Academy of Sciences in Novosibirsk in the late Brezhnev era, a group centered around the sociologist Tat'iana Zaslavskaia. Beginning in the late Brezhnev era and then increasingly publicly after Gorbachev's rise to power, Zaslavskaia and her colleagues put forward an alternative analysis of the problems of declining work discipline and economic stagnation in the USSR—an analysis that came to be known as the "human factor" approach to economic reform. In a series of prominent articles in Soviet academic journals, Zaslavskaia argued that the overwhelming centralization of Soviet economic life that developed under Stalin and became ossified under Brezhnev had acted to prevent the expression of an otherwise natural enthusiasm for labor under socialist conditions. As she put it: "A great deal depends on the worker's recognition of his involvement in the common cause, on the degree of his subjective 'involvement' in the production process, on the level of his identification with the collective and with the contents of his job."[20] The more the worker identified with his or her work, Zaslavskaia argued, the greater "the fulfillment of plans and norms and the effectiveness of utilization of labor time."[21] Under Brezhnev, however, owing to the lack of any correspondence between a worker's effort and his or her material reward, "people with a high skill level and the ability to do excellent work did not want to work with maximum intensity and preferred the 'quiet life.' "[22]

However, Zaslavskaia's analysis did not lead her to accept, as did Shmelev, the idea of unemployment as a solution to problems of worker idleness. She argued that a cultural transformation in the USSR, bringing about a widespread enthusiasm for labor, was the only answer:

"There is but one truly promising solution: the inculcation of all categories of workers with the capacity for self-monitoring based on a high degree of professionalism, personal dignity, pride in the excellent quality of job performance, and an aversion to careless work."[23] Zaslavskaia remained somewhat vague on the issue of how this "inculcation" would take place; her assumption seemed to be that the mere experience of taking part in a "radically reformed" socialism in the USSR would provide enough impetus for workers to change their cultural attitudes toward labor: "The crucial conditions for accelerating social development are activation of the human factor, fuller and more effective use of the individual's labor and intellectual potential, and reawakening the creative energy of the masses and channeling it into the mainstream of social interests."[24] To create the preconditions for such a "reawakening" might, Zaslavskaia admitted, require a short-term concentration on narrow economic goals—weeding out truly lazy or incompetent workers, changing the system of incentives to reward hard work more effectively, and so on. But this "subordination of social goals to economic goals" could be justified "only for short periods of time, limited to one or two five-year periods. Over the longer term, the principal goal of the development of socialism becomes the creation of a more progressive system of social relations that will ensure the all-round development of the individual, the realization of personal aptitudes, and the renumeration of representatives of all social groups in accordance with work performed."[25] Thus, in rejecting Stalinist centralism without advocating any new centralism to take its place, Zaslavskaia implicitly held out the possibility that in a reformed USSR, for the first time in history, a truly self-actualized and self-disciplined working class would emerge. The Soviet Union would then have achieved a "qualitatively new level of development of social relations, and consequently, of man himself."[26]

In sum, Zaslavskaia's articles painted a heroically optimistic picture of the state of the Soviet economy and the potential efficiency of the Soviet workforce under the auspices of reformed Leninism. Under Brezhnev, the lazy had been rewarded more than those who took the initiative; reversing this equation would unleash a surge of enthusiastic, high-quality work that, combined with the high technology now available in the USSR, would begin to transform the Soviet Union into a qualitatively new type of social order. Upon closer examination, there was little in Zaslavskaia's work to indicate concretely what institutional changes the party should implement to change the Brezhnevian incentive structure. She called for increased penalties for violations of labor discipline but stopped short of advocating the ultimate penalty of unemployment; she called for greater pay differentials to reward skilled and high-quality labor but stopped short of recommending a free labor market, insisting on the need to "channel" individual interests to serve social needs. The one consistent policy implication of her analysis was a negative one: the overcentralized system of economic planning inherited from Stalin must be destroyed before the "human factor" could be activated and socialism's unique potential to "fully reveal man's abilities and develop his creative activity as an actor in social life" could be realized.[27]

Perestroika and Charismatic-Rational Time

The implications of the "human factor" approach to the reform of Leninism were concretized in the new slogan and strategy Gorbachev introduced in a decisive way in January 1987—*perestroika*, or restructuring.[28] The general secretary's call for the *perestroika* of existing Soviet institutions marked Gorbachev's final break with Andropovism and his emergence as a full-fledged innovator in the tradition of Marx, Lenin, and Stalin. Having realized the ultimate unworkability of right, left, or center strategies in the post-Stalin era for realizing the potential of a fully socialist cultural transformation in the USSR, Gorbachev drew the logical conclusion: Stalin's socioeconomic institutionalization of Leninism was itself to blame for the dead end in which the movement found itself. The natural tendency of socialism, after all, was to be dynamic, to outstrip time: something in the existing institutional structure of the Soviet economy and society must therefore be acting as a "braking mechanism" on this process.[29] Before an acceleration of Soviet socioeconomic progress could be achieved, therefore, it would be necessary in effect to begin all over again from 1928 and "restructure" the Soviet economy along new lines. This would be possible, moreover, without merely substituting a new set of mechanisms of central control for the old centralized planning system, because, as Zaslavskaia's analysis had insisted, Soviet society had "matured" to the extent that a more dynamic socialist culture would emerge naturally as Stalinism itself was destroyed.

Perestroika, then, was to be the cultural equivalent of Lenin's "professional revolutionary" seizure of state power in Russia or Stalin's combination of "American efficiency and Russian revolutionary sweep" in Soviet socioeconomic institutions. It represented a practical and disciplined, yet revolutionary, movement to introduce a fourth and final period of development in Marxism-Leninism, one based on a continual dynamic internalization of norms of "revolutionary discipline" on a mass basis; in this sense it was "in its Bolshevik daring and in its humane social thrust . . . a direct sequel to the great accomplishments started by the Leninist Party in the October days of 1917."[30] As Gorbachev rhapsodized: "The success of perestroika will be the final argument in the historical dispute as to which system is more consistent with the interests of the people. Rid of the features that appeared in extreme conditions, the image of the Soviet Union will gain a new attractiveness and will become the living embodiment of the advantages that are inherent in the socialist system. The ideals of socialism will gain fresh impetus."[31] The stakes were clear: a successful implementing of *perestroika* would not only reinfuse socialism with "fresh impetus" but would provide the "final argument in the historical dispute" between socialism and capitalism. As Gorbachev never tired of insisting, then, *perestroika* in its conception did not involve any rejection whatsoever of socialist values. It was meant instead to bring "more socialism" into the Soviet way of life and to cleanse the deformations of socialism that took place during the "era of stagnation."[32]

However, formal intentions are not the same as practical outcomes. Operating on the theory, articulated by Zaslavskaia and others, that a "mature" Soviet culture awaited only the destruction of overcentralized control by the party-state over socioeconomic life to blossom into a disciplined and dynamic force for economic development, Gorbachev defined *perestroika* in almost purely negative terms. *Perestroika* in practice meant the disciplined revolutionary destruction of Stalinism, not the construction of any particular institutions to take Stalinism's place. The constructive side of *perestroika*, the creation of new forms of socialist life, was to be left to the popular initiative of the masses; any departure from this principle was held to be a reversion to Brezhnevian or Stalinist ways of thinking. If the theory of Soviet society under which Gorbachev was operating had been correct, a new culture of disciplined revolutionary activity in everyday life should have begun to emerge in the USSR—not all at once, as Khrushchev had imagined, but gradually, and then more and more dynamically. New "plans" for revitalizing society from the center, in this view, would only slow down this process.[33]

Accordingly, the three major campaigns associated with *perestroika*— for *glasnost'* (publicity), for *demokratizatsiia* (democratization), and for "new thinking" in Soviet foreign policy—were each directed mainly *against* the existing order rather than toward the creation of a new order. *Glasnost'*, which encouraged the open exposure in the Soviet press of the horrors of Stalinism, of current party corruption, of Soviet social problems, and, most consequentially of all, of Soviet nationality disputes, was designed to discredit for all time the centralized form of Leninism that had taken shape under Stalin and fossilized under Brezhnev—a task it performed with unparalleled effectiveness. *Demokratizatsiia*, in its original implementation, was designed to elicit mass pressure against the continuing tenure of die-hard Brezhnevites in party and state positions and to provide a political opening for grassroots innovation. Even the dramatic creation of the legal foundations for a more independent Supreme Soviet in the fall of 1988 was intended above all to provide an alternative power base for Gorbachev in his struggle to destroy the existing party and state apparatus, rather than as an attempt to establish Western norms of liberal parliamentarianism in the USSR.[34] And the "new thinking" in Soviet foreign policy amounted to an absolute rejection of the fundamentals of Brezhnevian norms of "socialist internationalism," most importantly in relations with the regimes of Eastern Europe, without any clear articulation of new principles according to which the Soviet use of its military and economic power abroad might be justified.

Glasnost', *demokratizatsiia*, and "new thinking" thus did not add up to a concrete vision of a "restructured" Soviet Union. Nor were they supposed to. They were designed to unleash popular initiative for the destruction of the corrupt, bureaucratized USSR that had emerged under Brezhnev and to make the process of *perestroika* within Leninist institutions "irreversible." In this task the three campaigns succeeded perhaps beyond Gorbachev's expectations. Within three years after the full-scale implementation of *perestroika* in 1987, every existing Leninist institution at home was under attack, and abroad the former Soviet bloc had crumbled into nothingness.

However, this was obviously not success in the form Gorbachev had originally imagined it. The Soviet economic situation, rather than undergoing steady improvement with the unleashing of mass initiative from below, simply continued to deteriorate as the Stalinist planning system lost whatever coherence it had still possessed in the Brezhnev era. New political openness had not merely relativized the role of the party apparatus but had led to the formation of powerful groups actively hostile to every aspect of party rule in the USSR, including fascist groups such as *Pamyat'*, liberal capitalist groups such as the Democratic

Union, and national separatist movements in practically every Soviet republic. The rejection of the "Brezhnev Doctrine" in Eastern Europe had not created a new multipolarity in international relations. It had resulted in the outright defection of the most strategically important Soviet allies—Poland, East Germany, Hungary, and Czechoslovakia—to the capitalist camp.

Worse yet for Gorbachev, *perestroika* itself, having already accomplished much of its destructive mission, had by 1989 begun to run out of steam. Already, as in earlier periods of revolutionary fragmentation in the history of the Marxist-Leninist movement, right, left, and center interpretations of *perestroika* itself had begun to emerge. The left interpretation, true to its charismatic essence, argued for a rapid destruction of the remnants of Leninism in the USSR, which would bring about a quick, miraculous return to "normal" life under the auspices of the market; such an antiparty interpretation of *perestroika* was at the core of the political platform of Boris Yeltsin. The analytically rational-legal, right interpretation of *perestroika* saw reform as a gradual, long-term process of the institutionalization of norms of liberal democracy in the USSR, with no possibility of any miraculous resolution of Soviet economic or political problems; this interpretation was argued most forcefully by Andrei Sakharov before his death on December 4, 1989. Finally, an orthodox center interpretation of *perestroika* itself began to emerge, emphasizing the preservation of Leninist party rule in any "reformed" Soviet Union and calling for crackdowns on the left and right alike, which were seen (indeed correctly) as objectively anti-Leninist. This theme was most consistently expounded by Yegor Ligachev.[35]

Gorbachev responded to the new splits in the *perestroika* movement in typical fashion. Rejecting, like Ligachev, both Sakharov's democratic right and Yeltsin's populist-charismatic left but at the same time painting Ligachev himself as symbolizing nothing more than a return to the discredited policies of Brezhnevian orthodoxy (which would in fact have been the likely outcome of a Ligachev victory), Gorbachev in early 1990 called for the continuation of *perestroika*. But *perestroika* was now to be directed not only against Stalinist socioeconomic centralism but against Leninist *party* centralism as well. In pushing for the repeal of Article 6 of the Soviet Constitution, which guaranteed the party's "leading role" in Soviet society, Gorbachev did not reject the notion of the vanguard party per se. But in Gorbachev's claim that the party would henceforth have to "earn" its leading role through concrete deeds, and in competition with other political forces, an implicit departure from Lenin's political thought began to manifest itself—for the implication of this argument was that the party could, in practice, become a "brake" on socialist progress, rather than being, as in Lenin's work, the ultimate *guarantor* of the socialist nature of socioeconomic and cultural development. Gorbachev's logic had led him to the conclusion that revolutionary Leninism might in principle require the destruction of the Leninist party itself.

Thus, although Gorbachev formally remained within the Marxist tradition even as late as 1990, substantively he had moved to a theoretical position that precluded any practical institutionalization of the Marxist ideal in political life. In his effort to put into practice a vision of cultural communism similar to Khrushchev's without becoming an unrealistic, Khrushchevian sort of leftist, Gorbachev had called for the disciplined dismantling of Stalinism. But this had resulted in his becoming like an earlier leftist, Leon Trotsky—railing against the Stalinist bureaucracy without any concrete economic alternative besides a vague faith in the enthusiasm of the masses. Aware that he had once again reached an impasse, Gorbachev in 1990 began to associate himself with the disciplined dismantling of Leninism itself, acquiescing in, and even seeming at times to encourage, the emergence of a fundamentally anti-Leninist political coalition centered around Boris Yeltsin in Russia and the de facto breakup of the rest of the Soviet Union along the lines of its constituent republics. But the continuation of this political line would very quickly leave him in much the same position as the original Marxist leftist, Rosa Luxemburg: a pure revolutionary romantic, believing absolutely in the creative power of the masses, unable to countenance in principle any concrete institutionalization of revolutionary politics that might stifle this creativity, and therefore doomed to be defeated by others who had no such scruples.

By the fall of 1990, *perestroika* had led to an incipient loss of control by the CPSU leadership over the Soviet Union's political future. Gorbachev's attempt to maintain his position by compromising between pro-market forces and national movements on the one hand and defenders of the old Stalinist order on the other could not be sustained. After Gorbachev's rejection of the five-hundred-days plan, those who now considered themselves true "reformers" deserted him en masse. In response, Gorbachev began to rely on hard-line elements within the party leadership—Kryuchkov, the KGB chief; Pugo, the head of the Interior Ministry; and Yazov, the head of the Red Army. These men had little patience for Gorbachev's continuing efforts to defend the *pere-*

stroika movement in the face of mounting political and economic disaster. Yet, given the utter bankruptcy of the Marxist-Leninist legacy, they could formulate no positive program of their own. Faced with Gorbachev's decision in the summer of 1991 to sign a new Union Treaty giving de facto sovereignty to all fifteen Soviet republics, Kryuchkov, Pugo, and Yazov in desperation mounted a coup against him. The "August coup" fell apart three days later, for all intents and purposes sealing the fate of the Marxist-Leninist revolutionary experiment.

Essentially, under Gorbachev, Leninism self-destructed. Yet this was, in one sense, a fitting conclusion to the 150-year historical progression linking Marx to Gorbachev. Marx had begun that progression by calling for the political establishment of communism as "the *real* movement which abolishes the existing state of things," a movement disciplined by its scientific analysis of time-bound history but uncompromising in its ultimate rejection of "bourgeois" time. Lenin had made communism a political reality for the "professional revolutionaries" of the Bolshevik Party in 1917; Stalin had made communism a socioeconomic reality for the Stakhanovites and revolutionary managers who had heroically overfulfilled their norms and plan targets in the early years of Soviet industrialization. But Marx's vision of communism as a permanent *cultural* ideal—as a principle of day-to-day life for the masses—was inherently unachievable, at least in a world still governed inexorably by time constraints. With the failure of Khrushchev's charismatic promises and Brezhnev's defense of orthodoxy as strategies for the creation of socialist culture, Soviet socialism after Stalin's death had inevitably itself become part of the "existing state of things." Gorbachev's *perestroika* abolished this state of things, and did so in a remarkably disciplined and revolutionary manner. In working to destroy Stalin's and Lenin's historical legacy, he was thus a faithful Marxist.

CONCLUSION

The collapse of the Soviet Union in 1991 can be seen as the analytic endpoint of a 150-year revolutionary experiment in the building of a new type of society based on the charismatic-rational conception of time. Remarkably, in the territory of the old Russian empire, an entire political and socioeconomic system had been built in response to Marx and Engels's call for a "revolutionary movement against the existing social and political order of things."[1] With these words, Marx had successfully transformed Hegel's original theoretical synthesis of charismatic and rational time from an essentially tragic philosophical doctrine into an inspiration for revolutionary action—one based on the core idea that successful time transcendence demands a healthy respect for the constraining power of history. This theoretical core of Marx's work was translated by Lenin into the basis for a new type of political organization and by Stalin into the basis for a new type of socioeconomic order. Only the final stage of this revolutionary experiment—Gorbachev's attempt to create a mass culture of disciplined time transcendence through the campaign for *perestroika*—turned out to be unrealizable in practice.

As we have seen, an understanding of the original worldview of the theoretical founders of the Marxist revolutionary movement is crucial to the interpretation of the whole course of Marxist and Leninist history from 1848 on. In fact, the continual process of charismatic-rational time's synthesis and subsequent decay in the theoretical, political, and socioeconomic cycles of its development is difficult to account for any other way.

To summarize: in each cycle, a crucial innovation in the theory and practice of charismatic-rational time by a founder—Marx in the theoretical cycle, Lenin in the political cycle, and Stalin in the socioeconomic cycle—was followed, after the founder's death, by a loss of revolutionary momentum and a split of the movement into three groups: a left group stressing a more fully charismatic interpretation of Marxism, a right group adopting a more fully rational-legal, procedural Marxism, and a center group more dedicated to attacking left and right "deviations" from orthodoxy than to fighting for further revolutionary advance. If left Marxism was analytically charismatic in stressing the miraculous achievement of revolutionary goals in a short period, and right Marxism was analytically rational-legal in stressing the empirical laws governing social evolution, centrist Marxism could best be described as neotraditional—based on a quasi-theological understanding of the "ideology" as infallible dogma, with the "orthodox" leadership seen as the sole legitimate interpreter of that dogma. The splits between these three types of Marxism, in each cycle, led to the decline of the movement's revolutionary momentum, until a new synthesis of time transcendence and time discipline in action could be found.

The basic outline of this process is represented in Tables 1 and 2. Such a consistent pattern shows that the worldview developed in Marx's works had a crucial impact on the words and deeds of those who followed Marx's teachings in the Second International and within the Soviet leadership after 1917. But understanding this pattern depends crucially on an analysis of the original Marxian belief system—and especially its orientation toward time. If this analysis is correct, the failure of Western Sovietology to take Marx's theoretical work seriously in analyzing the later actions of the CPSU leadership has obscured the particular developmental features of a remarkable revolutionary regime. At the same time, it is clear that without a continual reliance on terror to induce artificially a sense of revolutionary struggle, this regime type would inevitably become corrupted and begin to disintegrate.

Marxism, Leninism, and Stalinism: An Assessment

It is tempting to conclude this historical analysis of the Marxist-Leninist-Stalinist experiment simply by proclaiming it a failure. But

Table I. *Cycles in the History of Marxism and Leninism*

	Revolutionary Innovator (charismatic-rational)		
	Left (charismatic)	Center (neotraditional)	Right (rational-legal)
Theoretical cycle		Marx ↓	
	Luxemburg	Kautsky	Bernstein
Political cycle		Lenin ↓	
	Trotsky	Zinoviev	Bukharin
Socioeconomic cycle		Stalin ↓	
	Khrushchev	Brezhnev	Malenkov
Cultural cycle		Gorbachev	

this would be too hasty a judgment. All regimes and ways of life are, in fact, temporary; regime "success"—to the extent that such a term can be meaningfully defined at all—should not be equated with mere longevity. The French revolutionary regime of 1789, even if one includes the period of the rule of the Directory, lasted for no more than ten years before Napoleon's coup, yet few claim that it was a "failed" revolution. The Ottoman Empire, by contrast, sustained itself for centuries in the role of a corrupt, unstable "sick man of Europe," but no one counts this as evidence of the empire's historical "success."

Rather, it is more instructive to judge the history of the USSR in terms of the original mission of its founders—even if one ultimately rejects that mission on ethical or practical grounds. What, when all is said and done, was the Marxist movement supposed to accomplish? As this study has shown, the men and women who debated how to make theoretical Marxism work in political practice, who fought for the Bolshevik Party in 1917–20, and who took part voluntarily in the cause of "socialist construction" during the first years of Stalinism, appear to have been motivated by a sincere desire to realize in practice Marx's vision of the creation of a communist society. Communism itself was to be understood not in "bourgeois Utopian" terms, as an ideal to be put off to the distant future, but as a "movement society" in which everyday life would become dynamic and revolutionary.[2] At the same time, commu-

Table 2. *Key Characteristics of Marxist and Leninist Factions*

	Revolutionary Innovator	Left	Center	Right
Ideological orientation	argues for new institutional and cultural synthesis of revolution and discipline	argues for achievement of full communism as rapidly as possible	defends fidelity to principles and institutions of the previous revolutionary innovator(s)	argues for slow evolution toward socialism within rational constraints
Domestic policy strategy	allies with centrists in defense of orthodoxy, then launches new revolution	calls for spontaneous mass campaigns against capitalism and "bureaucracy"	attacks "deviations" from orthodoxy and attempts to inculcate socialist values in society	advocates reform of material incentives to improve productivity
Foreign policy strategy	breaks with centrist internationalism to argue for feasibility of revolution in one country	calls for global revolution launched in many countries at once	calls for global revolution directed by the orthodox center	defends "legal" boundaries of existing socialist state
Outcome	creation of new revolutionary institutions, or collapse	marginalization by party elite; exile and/or assassination	stagnation and corruption of party bureaucracy	economic success at the expense of revolution

nist society was supposed to represent the pinnacle of scientific and technological achievement and economic efficiency. Finally, communism was meant to provide for universal personal fulfillment through work; it was a society in which unfulfilling labor would be reduced to a minimum while "self-actualizing" labor would become "life's prime want." In such a society, when the cultural victory of the communist ideal was complete, there would be no need for political or economic coercion of any kind, and goods would be provided "from each according to his ability, to each according to his needs."

Obviously, the Soviet experiment fell far short of reaching this goal. Yet, in retrospect, it is quite remarkable how long the vision of communism was sustained as a basis for policy in the USSR. Moreover, since the vision was one of continual activity, rather than of stasis, progress toward the ultimate goal was, in periods of successful revolutionary advance, tantamount to realizing the goal itself. After all, Marx wrote that communism should be understood not as a "state of affairs to be established" but as "the real movement which abolishes the present state of things." By this standard, the Bolshevik Party under Lenin made communism a reality for many party members in the years of the revolution and the Civil War. In the same way, for those who believed in the cause, the Soviet Union was genuinely realizing communism during the years when Stalinist charismatic-rational socioeconomic institutions were being successfully constructed.

Such an outcome was certainly not what Marx and Engels had in mind when they wrote the *Communist Manifesto* in 1848. But in crucial respects, the historical development of Leninism and Stalinism followed the basic outlines of Marx's vision of revolutionary socialism. Both Marx and Lenin had understood and explicitly argued that before "full communism" could be reached, the revolutionary movement had to pass through several intermediate stages: first, the establishment of a political "revolutionary dictatorship of the proletariat," then a lengthy battle with residual capitalist elements in the management of socioeconomic affairs, and then finally a struggle against "survivals of capitalism"— and indeed, against "survivals" of the whole period of human "prehistory"—in mass culture. Consistent Marxism did not require the simultaneous resolution of all these tasks—as did anarchism, for example—but only that each task be approached in a consistently revolutionary and time-disciplined manner, neither attempting impossible leaps forward nor reconciling oneself to the constraining power of time. In fact, it was precisely when the leaders of the movement began to think of the ultimate goal of communism as separate from the movement itself— that is, when they began to think of time in the abstract, linear terms characteristic of Western liberal thought—that splits between left and right factions and the subsequent decay of the movement began. For this disease, merely the preservation of orthodoxy by the center was no cure; only by further revolutionary advance could the sense of time transcendence within time be maintained.

What is absolutely unprecedented about the Leninist experience, however, was that the Bolshevik Party elite managed successfully to industrialize a traditional agrarian economy relying primarily on the internal resources of the USSR while simultaneously maintaining a charismatic emphasis on collapsing time within its major economic institutions. Stalin's achievement in creating an "economy of a new type" in the USSR—one based on the continual smashing of "bourgeois" time norms—must be understood in this light. Stalin's economic innovations allowed the shock worker, Stakhanovite, or revolutionary manager to be heroic within the confines of mundane economic life—to *routinely smash temporal constraints.*

This having been stated, however, it is crucial to stress the other side of the Leninist experience, what from a Western perspective (and certainly from my own point of view) is its dark side. That is, one must emphasize—while not forgetting the social upheaval involved in industrialization under the auspices of liberal capitalism in the West—the extremely high degree of violence utilized in the implementation of Soviet political and economic institutions. This violence, I think, is intimately connected to the failure of Marx's conception of time as a *cultural* ideal, despite its temporary success in approaching theoretical, political, and even socioeconomic questions.

The cultural ideal of Marxism—encapsulated in Marx's dictum that labor under communism will become "not a means to life, but life's prime want"—inspired a standard of proper revolutionary behavior that very few people could live up to in practice, and only then for very short periods of time. The "proper" Bolshevik was one who spent hours at party meetings, spent more hours listening to and acting upon the complaints of his subordinates, and then went home to study political documents and raise his technical skills.[3] The "proper" revolutionary worker was one who consistently overfulfilled her production norms, donated her Saturdays to the cause of communist construction, and used her remaining time to better herself through attendance at cultural events or adult school.[4] By these standards, anybody who simply wanted a formal break from activity—which meant, in practice, almost everybody—could be seen as "objectively counterrevolutionary." Under Stalin, this sort of reasoning could have murderous consequences, as was evidenced by the mass executions in the Great Purges of those members of the party elite who had advocated an analytically rational approach to the economics of socialism in the twenties.

However, Stalin's terror of 1936–38, despite its scope, affected primarily the Bolshevik elite. On the mass societal level, the cultural violence exercised in the name of charismatic impersonalism was directed against those who participated in traditional ways of life based on the natural cycles of the agrarian economy—that is, the overwhelming peasant majority of the Soviet population. The collectivization of agriculture, which was the most decisive break with prerevolutionary culture effected by the Stalinist regime, resulted in the deaths of millions of people and destroyed the basis of traditional village life in the USSR—

all for the sake of implementing an ultimately unworkable alternative ideal of disciplined revolutionary economic activity.

Given this extreme degree of violence, the question becomes how Stalinist charismatic rationalism was successfully established in Soviet society at all, especially in the years before the Soviet victory in World War II provided the regime with a certain degree of traditional nationalist legitimacy. Here we should focus, following the work of Sheila Fitzpatrick, on the one group that benefited enormously from Stalin's methods of industrialization: that group of Soviet workers, one or at most two generations removed from the countryside, who joined the party during the twenties and took seriously Stalin's call to smash the resistance of "bourgeois wreckers" in town and country during the First Five-Year Plan.[5] These individuals identified with the party's ethos of simultaneous heroism and efficiency and cherished their status as the proletarian foundation of Soviet power. Through the institutionalization of five-year planning and shock work, and through the rapid promotions of those workers who adapted themselves to the extreme tempos of Stalinist production, Stalin created an incentive structure that produced at least a temporary realization of the Marxist cultural ideal among a significant minority of Soviet workers. Had the allegiance of these workers not been won, Stalinism could never have endured for as long as it did. Indeed, it was precisely this group that formed the core of the party elite that would rule the Soviet Union until the rise of Gorbachev.[6]

But over time, the commitment of even a minority of Soviet workers and managers to the struggle to compress time in economic production began to waver. And here a crucial difference between the decisively rational conception of time institutionalized in liberal capitalist economic life and the Leninist conception of time must be pointed out. The ethos of capitalism, as was argued in Chapter 1, is based on the acceptance of abstract time as itself sacred. The cultural ideal of capitalist work thus tends over time to work *against* heroic attempts to defy time's dominion. This, in turn, means that the elites who organized capitalist industrialization never felt compelled to ignore in principle, as the party elite under Stalin did, the necessity of *time off*, both from labor and from leisure-time efforts at self-improvement. To be sure, the good Puritan was supposed to keep sleep and recreation to an efficient minimum, but to indulge in round-the-clock activity at the cost of one's future time efficiency was also seen as sinful. "Early to bed, early to rise, makes a man healthy, wealthy and wise," Benjamin Franklin admonished— thereby affirming the importance not only of going to bed but even of doing so as early as possible so as to get a fresh start in the morning.

This is not to deny that the Protestant ideal of what constituted "sufficient" sleep for productive work—often no more than a few hours a night—could be quite coercive when enforced on reluctant factory workers. But, at least potentially, the modern idea of time as abstract and divisible contained the seeds of a new, nontraditional means of organizing the social division of work time and free time that, in principle, need not involve draconian labor discipline. As workers gained political power in the West, they could effectively bargain for a more reasonable balance of work time, free time, and sleep time and do so without challenging the fundamental cultural norms of time discipline crucial to the ethos of capitalism.[7] Ironically, the passage of the Ten Hours' Bill in England, which Marx saw as evidence of the impending revolutionary destruction of the capitalist time sense, was in fact a symbol of the cultural victory of the rational conception of time among the capitalist working class.

No such possibility of mass cultural reconciliation of the Soviet population to the temporal demands of charismatic rationalism as practiced in Stalinist economic institutions was, however, possible—at least, barring the actual emergence in Soviet society of a "new socialist man" who would never need time off from disciplined time transcendence.[8] The regime's rewarding of the overfulfillment, rather than the mere fulfillment, of plan targets and work norms, along with its efforts to organize a "rational" use of leisure time, did produce a *formal* compliance with the Marxist cultural ideal during the early years of Stalinist industrialization. But very quickly, an *informal* compromise with time in everyday life had to be made, if only to allow for periods of recuperation between revolutionary economic and cultural assaults.

The basis of this compromise in Soviet industry was the increasing acceptance, by the Soviet elite and Soviet society alike, of the practice of *storming*—a periodic rush to meet plan targets, followed by a more or less prolonged cessation of effective work activity.[9] Soviet workers and managers proved unable to perform like Stakhanov every day—or even most of the year, for that matter. Instead, they made a tacit agreement with Soviet central planners to act like Stakhanovites at least once in a while—at the end of the monthly and yearly planning periods. Storming looked enough like revolutionary time discipline to satisfy the minimum demands of the party elite, and it felt enough like a traditional work cycle to satisfy a significant sector of the Soviet working class, so recently removed from rural life. Hence, despite periodic efforts by the party leadership to uproot this practice in favor of *"ritmichnost'"* (a rhythmical pace) in economic activity—efforts that, under Stalin, included the severest possible penalties for violations of work discipline—storming rapidly became the informal norm of Soviet work life.[10]

It is important to point out that the nearly universal spread of storming behavior in Soviet factories represented a real cultural change from the more purely traditional patterns of work and leisure characteristic of prerevolutionary Russian peasant society. No longer were the signals for intensifying or relaxing work activity given by nature, as in the seasonal progression of intense harvest labor followed by winter rest. Instead, these signals were formulated consciously by the party-state leadership through its central planning apparatus. This meant that the cycles of labor and leisure characteristic of storming could be, within limits, sped up artificially by the center. The traditional economy, based on the concrete cycles of the cosmos, could by its very nature never exceed one harvest period a year. The First Five-Year Plan, by contrast, achieved a constant "harvest" ethos in production for four years; later plans were less extreme but still met with significant success in raising the level of work intensity above traditional norms.

However, the cultural break with traditional modes of temporal economic organization effected by the process of Stalinist industrialization was not analogous to that brought about by liberal capitalist industrialization. Although Stalinist planning and work norms succeeded in periodically stimulating an artificial "harvest" ethos in Soviet economic production, this achievement was predicated on a reliance on continuous revolutionary economic advance that could not be sustained in a world still bound by the constraints of time. More or less continuous storming produced an industrial infrastructure in the USSR in "record time," but it was an infrastructure built on a shaky foundation. According to Stalin's own express desire, the building of the Soviet economy was a rush job—an attempt to do in ten years what capitalism had taken centuries to accomplish. That the effort succeeded at all is nearly miraculous; that the factories and equipment produced by charismatic planning and shock work were riddled with defects is hardly surprising.

In the long run, however, the Stalinist economic worldview, based as it was on the rejection of purely rational time, was ill suited to respond to the problems of time efficiency crucial to sustained economic growth. In fact, as Grossman has shown, those engineers and economists of the Stalin era who attempted to work out various "socialist" mechanisms for the economic valuation of time were vehemently criticized—and sometimes arrested and shot—for "being under the influence of bourgeois theories" and relying on "bourgeois concepts of scarcity."[11] The Stalinist assumption that a continual increase in the "production of the means of production" would obviate the need to deal with questions concerning the proper allocation of investment, the amortization of industrial equipment, or the setting of interest rates proved to be incorrect; but this, in turn, meant that heroic activity on the "labor front" became an increasingly ineffective response to Soviet economic decline. Steel factories in the Urals might be built by storming, but shock brigades were ill suited for the undertaking of the repair of existing machinery or for shifting over to the production of high technology or consumer goods.

Strikingly, however, the various "reforms" of the planning mechanism during the post-Stalin era never explicitly challenged the fundamental problem this situation raised: the impossibility of economic heroism in a "developed socialist" society. Relaxations of central control were still predicated on the notion that they would lead to a new period of voluntary, rather than coerced, plan and norm overfulfillment analogous to that of the First Five-Year Plan. The principle that the party should simply encourage the efficient *fulfillment* of industrial targets, while *actively discouraging overfulfillment*, was never introduced.[12]

The only way out of this dilemma within the confines of the Marxist-Leninist acceptance of a charismatic-rational time orientation—and to attack this time orientation would be to attack the principle legitimating Leninist party rule itself—would have been the actual cultural transformation of Soviet society in a charismatic-rational direction. In short, sustaining Soviet socioeconomic institutions in a "developed socialist" context required the empirical achievement of a communist society in the USSR where voluntary participation in labor would replace central direction, and where the need for continuing material incentives to stimulate heroic labor efforts would disappear.

This cultural problem for Marxism-Leninism had been only obliquely addressed by Marx; Lenin had recognized it clearly in his last years but failed to formulate a workable response to it. Stalin alone had provided a theoretical solution to the problem of demonstrating continual revolutionary progress in a battle against the dominion of time in human culture that, by its very nature, could never be won. This solution was encapsulated in his notion that "the closer we get to the final victory of socialism, the more enemies we shall have"—a doctrine that concretized the "continual abolishing of the present" central to Marx's ideal of communism as a continual process of revolutionary mass murder.

Unwilling to accept this solution—which threatened the destruction of even the most loyal supporters of Stalinist socioeconomic institutions—the party moved after Stalin's death to repudiate this aspect of Stalinism and the secret police terror based on it. But this left the cultural problem of Leninism—how to get Soviet society, on the mass level, to internalize norms of disciplined time transcendence—unresolved. Malenkov hinted at the possibility of emphasizing consumer goods production to encourage efficiency in economic construction. Khrushchev declared the party-state to be one "of the whole people" and attempted to engage the masses in continuous campaigns for overtaking bourgeois time constraints. Brezhnev simply concentrated on struggling against such right and left deviations from Stalinist socioeconomic practice and adopted Kautsky's posture of "revolutionary waiting"—a platform that depended ultimately on an overly optimistic assessment of the prospects of socialism in the Third World. Overall, the period between 1953 and 1985 was analogous to the two earlier "periods of stagnation" in the Marxist movement. Right, left, and center strategies alike had failed to preserve the core of the charismatic-rational conception of time—a sense of revolutionary forward movement that did not depend on "voluntaristic" wild leaps for which the social conditions were not "ripe."

The Gorbachev *perestroika* reforms of Soviet socialism were an attempt to break out of this impasse and to realize on the cultural level the fusion of revolutionary action with rational time discipline that Marx had created theoretically, Lenin politically, and Stalin in socioeconomic institutions. *Perestroika* was predicated on a particular set of assumptions about the nature of the Soviet Union in the post-Stalin era: that Soviet society had "matured" to the point where central direction was an impediment to further economic progress; that a decentralized USSR would, by releasing the hidden potential of the "human factor," lead to the desired norm of labor discipline, one based on voluntary enthusiastic labor as a form of free self-expression; and that the destruction of the centralized Stalinist economic system would result in there being not less but "more socialism" in Soviet life.

This set of assumptions turned out to be incorrect. Soviet society and culture had indeed undergone massive changes in the years during and after Stalin's rule, but not in the direction of a more complete internalization of socialist work and leisure norms, as Gorbachev and his idealistic advisers believed. Rather, several decades of Stalinist economic practice had broken down Soviet society into three major groups, each of which reacted differently to the failure of the promise of the socialist revolution in economic life.

First of all, more than a quarter of the Soviet population remained engaged in agricultural activity, still tied to traditional work rhythms based on natural seasonal cycles. However, this group was no longer bound by traditional forms of *political* authority. Decades of the collective farm system had effectively destroyed the family and village networks of prerevolutionary life in the countryside, and the system of rural party collectivism had failed to provide an effective substitute. The mass destruction of the kulaks in the thirties had deprived the countryside of its most energetic and successful farmers, and the migration of most skilled workers to the cities in the decades that followed left the rural areas of the USSR demoralized and stagnant.[13]

Approximately 65 percent of the Soviet population had, by 1985, become fully urbanized, but in urban conditions markedly different from those characteristic of the capitalist West.[14] The Stalinist system had created factories that were symbols of heroic revolutionary transformation rather than "schools of modernity"; Soviet workers in medium-sized cities remained by and large wholly unexposed to the time-disciplined, efficient management and labor practices characteristic of the developed capitalist countries. Larger Soviet cities themselves were at best ambiguous symbols of the desirability of industrial progress; they tended to be bizarre amalgams of the ancient and the contemporary, with both types of architecture decaying simultaneously. The urban population of the USSR, by and large, had been removed from traditional patterns of political and economic life in the countryside only to be subjected to a corrupt form of neotraditionalism in the city, which, as in the rural areas, had ultimately failed to become a legitimate substitute for older traditional ways of life.[15]

Finally, a small but significant part of Soviet society, particularly among the Soviet intelligentsia, had, through travel, education, and personal reflection, come to the conclusion that lies at the basis of socioeconomic organization under liberal capitalism: that abstract, linear time is inexorable and cannot be overcome by revolutionary action and that economic life should be predicated on using time efficiently rather than transcending it. They faced the unenviable problem of reconciling this intellectual conclusion to the reality that, as in the West, decades of organizational mobilization and coercion would probably be necessary to impose patterns of capitalist time discipline among the potentially hostile working and agrarian classes.

None of these groups were likely to respond to Gorbachev's decentralization of political and economic authority in the ways predicated by

Zaslavskaia and other theorists arguing for an increased reliance on the "human factor" in Soviet economic life. For the traditional rural population and the neotraditional urban workers and managers, the relaxation of central control simply stripped the Stalinist system of its last mechanisms for ensuring even minimal compliance with work norms and planning targets. The cooperatives—the one new arena that provided some incentive for independent economic initiative based on one's own internalized time discipline—were widely attacked by these groups as mere mechanisms for profiteering at the expense of the mass of the Soviet working population.[16] On the other hand, both liberal and nationalist intellectuals more or less consciously used the political opening provided by *glasnost'* and *demokratizatsiia* to push for the complete destruction of the Marxist-Leninist worldview and accompanying institutions. Meanwhile, the Soviet economic crisis continued to worsen.

It is reasonable to conclude, then, that the Stalinist economic system was inherently unreformable—not because of its reliance on central planning, which exists in various forms in liberal capitalist regimes as well, but because of its fundamental emphasis on mastery over time itself as the standard of economic success. Ultimately—as the Weberian approach adopted here, with its stress on the necessary routinization of charisma, implies—the force of time in economic life cannot be abolished or artificially compressed; the Stalinist economic system was thus bound to lose its coherence and effectiveness at some point. Gorbachev's *perestroika* destroyed the Soviet polity and economy more dramatically and more rapidly than would otherwise have been likely, but the ultimate disintegration of the Soviet system was, in any case, inevitable.

The Weberian Theory of Social Time: Concluding Remarks

That the formal conception of time contained within Marx's theorizing was critical in generating the patterns of development in the history of Marxism-Leninism from Marx to Gorbachev raises important questions for comparative research. The Weberian theory of time and social organization, used here to analyze the rise and fall of the Soviet system, may prove useful for understanding other cases of ideological, political, socioeconomic, and cultural development. Shared formal conceptions of time, it might be argued, are always in some sense at the root of social institutions, since without common agreement about the temporal order even the most minimal cooperation between human beings would be impossible in the absence of face-to-face interaction.[17] Indeed, in the foundation of large-scale bureaucratic political parties and legally enforced socioeconomic systems such as those characteristic of the modern industrialized world, explicit and detailed theories about the nature of human activity in time have often played a crucial role.

In analyzing the history of the rise and fall of a given regime, I would argue, understanding the ideological or religious definitions of time of its founders may be important for three reasons. First, the conceptions of time and eternity shared by state-building elites help to define the state's political identity, since the values and behaviors held to be "sacred" or time-transcendent by a given religion or ideology often provide the main criterion by which members of a new polity can be distinguished from nonmembers. In this sense, charismatic or time-transcending elements are always at the core of any novel "regime identity." Second, as we have seen, an elite's notions of how daily time use should be organized can directly affect the patterns of behavior institutionalized politically and socioeconomically in a given regime. Finally, important aspects of cultural change under the auspices of particular regimes can be understood as emerging out of the clash between elite attempts to impose notions of "proper" time use through formal social institutions and the informal mass resistance of various social groups attempting to preserve their habitual patterns of daily behavior.

The history of the USSR, from this perspective, provides a fascinating case study of what may actually be a more general evolutionary process.[18] Enormous work remains to be done, however, in tracing the development of other polities, economies, and cultures back to their origins in the worldviews of philosophical, ideological, or religious elites. For too long, the false dichotomy between "ideas" and "material interests" accepted by social scientists has simply precluded research into the historical links between specific idea systems and the institutionalized incentive structures that emerge out of them. Of course, the study of time conceptions is only one problem among many that could be pursued within this broader social scientific project. Hopefully, though, this study has made some small contribution toward opening up related lines of inquiry.

Finally, the study of time and Soviet institutions has important implications for the way we view the more purely rational time discipline characteristic of Western liberal capitalism, for liberal capitalism, too, can be analyzed as a distinctive regime type that has passed through definite theoretical, political, and socioeconomic stages of development and has produced an empirically observable cultural change in the time orientations of the populations living under it. Bearing this in mind, we should not come away from this analysis of the rise and fall of the Soviet experiment with a smug assurance that liberal capitalism has won a final victory in the "ideological battle" of history, or that history itself has in some sense come to an end.[19] Such beliefs have nothing to do with comparative social science and everything to do with the ideological self-understanding of liberal capitalist elites, who think that they have implemented political and economic procedures that are uniquely inherent in the unfolding of time itself and that these procedures will therefore last as long as time does.

Rather, the Soviet experience should be analyzed in order to gain insight into the sorts of stresses and strains that confront *all* regimes in their attempts to reconcile the problem of political legitimacy, which depends on a sense of the regime's connection with the realm of eternity, with the problem of material scarcity, which is inherent in social life governed by time. In this context, the collapse of the USSR forces us to reevaluate the temporal order typical of Western regimes in a new light. What is the nature of present-day liberal capitalism's claim to eternal legitimacy? And how might the realities of economic scarcity inherent in the time-boundedness of life on earth threaten that claim at some point in the future?

Weberian analysis does not pretend to provide final answers to questions like these. But such analysis *does* indicate where new social tensions are likely to arise in a given ideological and institutional setting. To the extent that the Weberian analysis of social time perception provided here is correct, a central problem for any regime is how to reconcile the contradictory demands of traditional, rational, and charismatic temporal outlooks—each of which represents a fundamental aspect of the human experience. People throughout history have wanted to feel connected to concrete cycles of generational renewal, to have their individual, linear life spans respected as a sphere for the expression of their personalities, and to experience a direct connection to something eternal, something that transcends the continuously corrosive force of time. The pure types of traditional, rational, and charismatic domination each absolutize a different element of this triad of human relationships to time. But even where formal institutional rules are strictly enforced by elites enjoying substantial legitimacy, actual social orders only approximate these ideal types. Whichever one of the three temporal principles a regime emphasizes, it is inevitably confronted with cultural demands for a reinfusion of the other two types of time into social life.

Sometimes, as in the case of the Soviet political and socioeconomic system, with its emphasis on continual disciplined time transcendence in everyday life, the worldview of a regime's elite and the institutions built upon it prove too inflexible to accommodate these cultural demands past a certain point, and the regime collapses. This sets the stage for the appearance of new charismatic principles by which mundane temporal life might once again be connected to the realm of eternity. The chaotic disorder left in the wake of the collapse of the Soviet Union, in this sense, represents a fertile environment for the generation of novel ideologies. Most of these will fail to find social support—but it remains possible that others will serve as the basis of unprecedented political and socioeconomic institutions. Thus new regime types—for good or for ill—may once again be generated on the territory of the former Soviet Union.

Notes

PREFACE

1. The best overall description of the Stalinist economic system in its fully developed form is given in Kornai, *The Socialist System*.

2. Indeed, some Soviet sociologists were claiming that this conflict had nearly been overcome during the reign of Brezhnev. As A. A. Gordon and E. V. Klopov put it: "At the level of development of our society, when the construction of communism has become a practical task and a realistic goal, the nonproductive life-activity of a man, his way of life and leisure time are as important for the development of personality as work in social production." Gordon and Klopov, *Chelovek Posle Raboty*, 9.

3. A similar argument has been made by Grossman, "Economics of Virtuous Haste."

4. See Jowitt, "Organizational Approach," 1183.

5. Jowitt, *The Leninist Response to National Dependency*, reprinted as "The Leninist Phenomenon," in idem, *New World Disorder*. Jowitt's neo-Weberian term for the principle of legitimacy institutionalized in the Leninist party is "charismatic impersonalism"; the phrase "charismatic-rational" used here is essentially synonymous.

6. There are by now a number of excellent works detailing the responses of Soviet social groups to Bolshevik policies governing time use in economic production. See, for example, Chase, *Workers, Society, and the Soviet State*; Chase and Siegelbaum, "Worktime and Industrialization"; Filtzer, *Soviet Workers and Stalinist Industrialization*; idem, *Soviet Workers and De-Stalinization*; idem, *Soviet Workers and the Collapse of Perestroika*; Siegelbaum, *Stakhanovism*; and idem, *Soviet State and Society*.

7. The most interesting comprehensive treatment of the influence of Marxist ideology on Soviet history is provided by Malia, *The Soviet Tragedy*. For the general role of Marxist-Leninist ideology in Soviet politics in the post-Stalin period, see, for example, Evans, *Soviet Marxism-Leninism*; Scanlan, *Marxism in the USSR*; and Thompson, *Ideology and Policy*. Finally, Walicki, *Marxism and the Leap to the Kingdom of Freedom*, deals with many of the same themes investigated here but arrives at rather different conclusions.

CHAPTER ONE

1. O'Malley, *Keeping Watch*; de Grazia, *Of Time, Work, and Leisure*, 82; Landes, *Revolution in Time*.

2. On the topic of time and society, see Adam, *Time and Social Theory*; Boorstin, *The Discoverers*; Fraser, *Voices of Time*; Giddens, *Constitution of Society*; Hall, *Dance of Life*; Rifkin, *Time Wars*, which includes an excellent bibliography; and Zerubavel, *Hidden Rhythms*.

A few accessible accounts of debates about time among natural scientists are those of Campbell, *Winston Churchill's Afternoon Nap*; Gould, *Time's Arrow, Time's Cycle*; and Hawking, *Brief History of Time*.

There are fewer works that directly deal with time as a political issue. See, however, Maier, "Politics of Time"; Gunnell, *Political Philosophy and Time*; Eliade, *Cosmos and History*; and Elias, *Time*. I have been especially influenced by the latter two works.

3. Elias, *Time*. Such generalized descriptions of traditional and modern society should be understood here as Weberian "ideal types"; that is, they are not meant to be precisely accurate descriptions of actual empirical societies, or of every person in a given society, but rather epistemological constructs that allow for broad comparison of social phenomena that would otherwise be lost in the unique characteristics each society possesses when studied individually. Thus, even in a decisively modern society such as that of the United States, the question "When are you *going* to finish your book manuscript?" can be answered with the word *eventually*, which is analytically traditional—tied to the flow of concrete *events*—rather than with a precise date, as would be more generally characteristic of the modern orientation toward time.

4. The way time is tied to concrete events in traditional cultures can be illustrated by a story related to me by an acquaintance who had just returned from a trip to the People's Republic of China in the mid-1980s. My acquaintance happened to be at a rural train station on the day that the Chinese government introduced daylight saving time for the first time. The posted train schedules had all been crossed out, and every arrival and departure time had been rewritten for one hour later. Apparently, the idea of "moving" time, without moving the events that mark time on a day-to-day basis, didn't translate culturally.

Conversely, the abstract quality of seconds and minutes can be seen in the way people in modern societies respond to the malfunctioning of a wristwatch. Rather than assume that seconds and minutes are now going more slowly than before, we conclude that our batteries are running low. In fact, we imagine that seconds and minutes would "go" at the same rate even if everybody's watches broke at once!

5. However, it should be emphasized that the distinction between cyclical and linear time, which is often seen as the fundamental distinction between the temporal sense of traditional and modern societies, is only loosely tied to the more important distinction between concrete and abstract time. Thus, one finds linear descriptions of concrete temporal events, such as the Old Testament, and abstract temporal cycles, such as the "fiscal year." As a rule, though, the cyclical metaphor for time will predominate in traditional cultures and the linear metaphor in modern ones. I would like to thank Russ Faeges for making this point clear to me. The simple equation of cyclical time with traditional time sense has also been persuasively criticized in Gunnell, *Political Philosophy and Time*, 110–12.

6. E. E. Evans-Pritchard, *The Nuer* (New York: Oxford University Press, 1982), originally published in 1940, p. 103.

7. See Gunnell, *Political Philosophy and Time*.

8. See Ariès, *Centuries of Childhood*.

9. Ariès, *Hour of Our Death*, 28.

10. Elkind, *The Hurried Child*; Mitford, *American Way of Death*.

11. Modern English has preserved this sense of the word *labor* in its usage as a synonym for childbirth.

12. For a fascinating discussion of the differences between the ancient and modern notions of labor and leisure, see Arendt, *The Human Condition*; also de Grazia, *Of Time, Work, and Leisure*.

13. See Eliade, *Cosmos and History*, for examples.

14. Schor, *The Overworked American*.

15. A very useful overview of work that addresses the issue of social time is provided by Adam, *Time and Social Theory*.

16. Inkeles and Smith, *Becoming Modern*.

17. Ibid., 160.

18. The phrase "schools of modernity" is taken from ibid., 154.

19. One peasant's response to the notion of maintaining modern time discipline ran as follows: "How could a man work like that, day after day, without being absent? Would he not die?" Quoted in Andrle, *Workers in Stalin's Russia*, 114.

20. Landes, *Revolution in Time*, 58.

21. Ibid., 71–72.

22. Ibid., 76; my emphasis.

23. Ibid., 77. Landes's response to his own objection here is that in western Europe, unlike in Japan, such a "traditional" mechanical clock was impossible because most clocks were inaccessible, at the tops of church towers, and therefore seasonal adjustment of clock faces was impractical. This, however, simply begs the question of why clocks were seen as so important to key medieval elites that they required such a central and prestigious placement as on church towers, even at the expense of a more flexible clock face more sensitive to traditional attitudes toward time. The complex interaction of merchant interests and changing Catholic theology that led to this result is suggestively analyzed in Le Goff, *Time, Work, and Culture*.

24. Thompson, "Time, Work-Discipline, and Industrial Capitalism," 73–74.

25. Ibid., 93. Thompson differs from the modernization theorists, however, in his Marxist hope that the teleology he describes will lead ultimately to a situation in which we can change our conception of time to "combine in a new synthesis elements of the old and of the new, finding an imagery based neither on the seasons nor upon the market but upon human occasions. Punctuality in working hours would express respect for one's fellow workmen. And unpurposive passing of time would be behavior which the culture approved" (96).

26. Hall, *Dance of Life*. For a general introduction to Hall's work, see *The Silent Language*.

27. Hall, *Dance of Life*, 16. The eight types of time Hall lists are sacred time, profane time, micro time, sync time, personal time, biological time, physical time, and metaphysical time.

28. Ibid., 5.

29. In fact, Hall reintroduces the equivalent of the traditional-modern distinction by referring later on to "polychronic" versus "monochronic" time perception, but he then fails to integrate this distinction with the framework he sets out in his first chapter.

30. Weber, *Economy and Society*, vols. 1 and 2.

31. Weber, "Social Psychology of the World Religions," 280.

32. Weber, *Economy and Society*, 1:215, 227.

33. Ibid., 244.

34. This is the same standard of legitimacy, in fact, that drives the investigations of modern science. Only those "laws" of matter that can be shown experimentally to be valid from the beginning of time are accepted as scientific; ad hoc explanations valid only for particular contingent outcomes are rigorously excluded. Arguments for politi-

cal legitimacy in modern societies also often take this form: market relations or democratic procedures are claimed to have the same timeless validity for human interaction as scientific laws for the behavior of matter.

35. The etymology of these words is fascinating: "appointments" exist only where a belief in abstract time allows one to locate one's activities at precise "points" on the "timeline"; "deadlines" originally meant the lines beyond which prison guards would shoot wandering prisoners.

36. The transcendence of ordinary time under charismatic domination should not be confused with stasis or peacefulness. In fact, the disappearance of regular time constraints (whether these are thought of as concrete or abstract) is usually a terrifying and disruptive occurrence, unless it is experienced within controlled settings such as ecstasy during religious rituals, drug-induced hallucination, or orgasm—all of which retain a certain potential for terror.

37. Weber, *Economy and Society*, 2:1117.

38. See also Arendt, *On Revolution*.

39. Eliade, *Sacred and the Profane*, 104.

40. Weber, *Economy and Society*, 2:1111–12.

41. On time in Judaism, see Eliade, *Cosmos and History*, 102–12; Zerubavel, *Hidden Rhythms*, 105–37; and Gunnell, *Political Philosophy and Time*, 54–71.

42. Nisbet, *Social Change and History*, 63–76.

43. Weber, *Protestant Ethic*.

44. Eliade, *Cosmos and History*.

45. Interestingly enough, Franklin also originated the idea of "daylight saving time"—perhaps the ultimate expression of the modern conception of time as abstract and disconnected from actual events.

46. For examples, see Thompson, "Time, Work-Discipline, and Industrial Capitalism," 89–96.

47. For an intriguing overview of the cultural changes brought about as a result of the standardization of rational time in western Europe, see Kern, *Culture of Time and Space*.

48. Quoted in Thrift, "Owners' Time and Own Time."

49. Perhaps the role of traditional time in academia should not be surprising. The Greek word for leisure, *schole*, is the basis for the English words "school" and "scholarship."

50. The relationship between time and gender roles is a crucial question that I will not be able to develop here. For an excellent introduction to the problem, see the essays in Forman and Sowton, *Taking Our Time*.

51. Only a few previous authors have examined the problem of time orientations in Marxist-Leninist ideology and the institutionalization of particular patterns of time use in the Soviet Union, focusing primarily on the early history of Leninism. See, for example, Chase and Siegelbaum, "Worktime and Industrialization"; Luke, *Ideology and Soviet Industrialization*; and Kaplan, *Bolshevik Ideology*. For an argument that Soviet work patterns derive from prerevolutionary Russian culture, see Leibovich, *The Russian Concept of Work*.

52. In a series of works written in the 1970s and 1980s, Jowitt proposed a neo-Weberian analysis of Soviet-type regimes that has greatly influenced the present study. Jowitt began by isolating what he termed the "genetic features" of Leninist rule—

those features that could not be changed without fundamentally altering the nature of the regime itself. These included above all the notion of the "leading role" of the Communist Party, to which all other social organizations were subordinated, and the claim that the party implemented its infallibly correct "general line" by combining a scientific analysis of society with the revolutionary insight of the vanguard elite.

In addition, Jowitt placed these core features of Leninism in comparative context by contrasting the nature of their legitimation to the form of legitimation characteristic of Western liberal capitalist states. Utilizing Weber's typology of rational-legal, traditional, and charismatic modes of domination, Jowitt argued that Leninist institutions were based on a novel form of "charismatic impersonalism" combining rational-legal proceduralism with a charismatic emphasis on the miraculous transcendence of ordinary political, economic, and cultural limitations. See the essays collected in Jowitt, *New World Disorder*.

53. Lowenthal, "Development vs. Utopia."

54. For obvious reasons, social scientists in the Soviet Union themselves were not generally in a position to examine the Marxist-Leninist conception of time critically. Nonetheless, certain Soviet philosophers have provided quite interesting discussions of the nature of time both in the history of Western thought and in the organization of society. See, especially, the work of Akhundov, *Conceptions of Space and Time*, and Trubnikov, *Vremia Chelovecheskogo Bytiia*.

55. Ken Jowitt, "Soviet Neotraditionalism," in *New World Disorder*.

CHAPTER TWO

1. Again, the ideal-typical nature of this concept of traditional politics must be stressed. Obviously, there have always been those in premodern settings who have questioned absolute conceptions of morality and truth; Thrasymachus in Plato's *Republic*, who argues with Socrates that "justice is the will of the stronger," is a famous example. However, in the context of Plato's dialogue, this remains an unphilosophical position; the conception of justice as a timeless quality of the soul wins the argument.

2. The philosophy of utilitarianism remains a dominant force today in Britain and the United States; normative versions of rational choice theory represent the most recent defense of the idea of utility maximization as the foundation of morality.

3. Although Kant's mature philosophical works embrace Newton's conception of time as the only possible basis for scientific understanding of reality, Kant's earliest writings appear to reject absolute time in favor of a fairly sophisticated argument for time's relativity to observers. See Al-Azm, *Kant's Theory of Time*, 11–14.

4. Kant, *Critique of Pure Reason*, 74–75.

5. Ibid., 75.

6. Kant, *Grounding*, 3.

7. Ibid., 39–44.

8. Kant shows his own cultural biases, for example, when he tries to argue that "indulgence in pleasure" is contrary to the universal moral law. See ibid., 31. Similar difficulties are encountered by an updating of Kantian moral theory, Rawls's *Theory of Justice*.

9. Kant, *Perpetual Peace*, in *Political Writings*, 116.

10. Kant, "Idea for a Universal History," in *Political Writings*, 53.

11. Ibid.

12. In fact, the project of reinterpreting Kant's "kingdom of ends" as an empirical goal, rather than a metaphysical construct, was common to most of the leading German philosophers after Kant's death. See Toews, *Hegelianism*, 30–48. On the influence of Hegel's social context on the development of his thought more generally, see Dickey, *Hegel*.

13. Hegel, *A General Introduction to the Philosophy of History*, published as *Reason in History*, 11.

14. Ibid., 12.

15. Ibid., 68.

16. Ibid., 71.

17. Hegel, *Phenomenology of Spirit*, 27.

18. Ibid.

19. Hegel, *Reason in History*, 87.

20. Ibid., 68.

21. Ibid., 91–92. It is interesting that in Hegel's symbolic account, it is an immortal god, not a human being, who is responsible for the initial entry of spirit onto the world stage. In a sense, Hegel's project here resembles that of Machiavelli, who was similarly concerned with the vitality of the state over time. But whereas Machiavelli saw human *virtu* as a countervailing force to the corruption brought about by Fortune, Hegel pits a kind of *supervirtu*—the spirit—against time itself, the "absolute negativity" that, unlike Fortune, cannot be beaten by masculine prowess or by republican virtue and against which merely human action is therefore impotent.

22. Ibid., 89.

23. Ibid., 95.

24. Hegel, *Philosophy of Right*.

25. This distinction between *moralitat* (morality) and *sittlichkeit* (ethical life) is fundamental to the argument of the *Philosophy of Right*; see ibid., 107–8.

26. For Hegel as a statist authoritarian, see Popper, *Open Society*; for Hegel as an evolutionary, generally liberal thinker, see Avineri, *Hegel's Theory*.

27. Hegel, *Reason in History*, 39.

28. Ibid., 40.

29. Ibid., 46, 43.

30. Ibid., 94–95.

31. Hegel, *Philosophy of Right*, 12–13.

32. Wilhelm Vatke, quoted in Toews, *Hegelianism*, 91.

33. Toews, *Hegelianism*, 95–140.

34. An interesting comparison and contrast of Hegelianism with Christianity is given in Tucker, *Philosophy and Myth*, 31–69.

35. Strauss, *Streitschriften*, quoted in McLellan, *Young Hegelians*, 3–4. The development of left, right, and center Hegelianism has been presented in rich detail by Toews, *Hegelianism*.

36. Toews, *Hegelianism*, 205.

1. Two recent authors have also taken time as their starting point in reexamining the work of Karl Marx: Postone, *Time, Labor, and Social Domination*, and Booth, "Economies of Time," 7–23.

2. The former argument is set out in Lovell, *From Marx to Lenin*. See also Fromm, *Marx's Concept of Man*. The latter position is argued in Kolakowski, "Marxism and Stalinism," in Tucker, *Stalinism*, 283–98.

3. The idea that Marx's theory is fundamentally dualistic—torn between its scientific and revolutionary components—has been argued by, among others, Gouldner, *The Two Marxisms*. However, my argument that Marx's work contains both a consistent core *and* a series of inconsistent arguments built upon that core is, to my knowledge, an original one.

4. See, for example, Avineri, *Social and Political Thought*; Thomas, *Karl Marx and the Anarchists*; and Tucker, *Philosophy and Myth*.

5. It should be noted that a whole school of Marxist thought absolutely rejects any attempt to discuss Marx's work in terms of "eternal values" and "morality." See, for example, Allen W. Wood, "Marx on Distributive Justice," in Cohen, Nagel, and Scanlon, *Marx, Justice, and History*. It is true that Marx often ridiculed such "bourgeois" concepts, stressing the class nature of all ethical precepts. However, my point here is not to label Marx a "moral" or "non-moral" thinker; rather, it is to place Marx in a philosophical tradition that believed in the power and value of modern empirical science while still holding certain qualities of human existence to be absolutely essential—irrespective of time or place—to human fulfillment. Even Wood, the most bitter critic of attempts to categorize Marx as a moral thinker, concedes that Marx believed in the absolute worth of what Wood terms the "non-moral goods" of "freedom, community, and self-actualization." See Wood, "Marx on Right and Justice," in Cohen, Nagel, and Scanlon, *Marx, Justice, and History*, 123.

6. MacGregor, *Communist Ideal*.

7. Thomas, *Karl Marx and the Anarchists*, 23.

8. Marx, "Theses on Feuerbach," in Marx and Engels, *Karl Marx Frederick Engels Collected Works*, vol. 5 (1976), p. 5. (This series will hereafter be cited as *MECW*.)

9. This point has been argued by Meyer in *Marxism*.

10. Marx, *Economic and Philosophical Manuscripts of 1844*, in *MECW*, vol. 3 (1975), p. 276 (hereafter cited as *1844 Manuscripts*). Emphasis here and throughout the notes to this chapter is that of Marx. For a detailed treatment of the concept of labor in the *1844 Manuscripts*, see Fromm, *Marx's Concept of Man*.

11. Marx, *1844 Manuscripts*, 276.

12. Marx and Engels, *The German Ideology*, in *MECW*, vol. 5 (1976), 31–32.

13. Marx, *1844 Manuscripts*, 305. This passage is especially significant in light of its placement immediately following a long, dense discussion of the question of historical creation in time, in which Marx argues that linear conceptions of history that extend backward infinitely toward some original creation neglect to "hold on to the *circular movement* sensuously perceptible in that progress, by which man repeats himself in procreation, *man* thus always remaining the subject," and are based on an unreasonable "abstract[ion] . . . from man and nature." Since man in effect creates himself through labor, charismatically negating the negation of time, purely linear conceptions

of history are mere abstractions from sensuous reality. Marx's philosophical rejection of the modern conception of time as an abstract, linear grid outside nature is quite clear here despite his rather difficult language.

14. Ibid., 274.

15. Marx and Engels, *The German Ideology*, 52–53.

16. Marx, *1844 Manuscripts*, 306.

17. Marx and Engels, *The German Ideology*, 49.

18. Marx, *1844 Manuscripts*, 305; Marx and Engels, *Communist Manifesto*, in *MECW*, vol. 6 (1976), p. 482; Marx, preface to *A Contribution to the Critique of Political Economy*, in *MECW*, vol. 29 (1987), p. 264.

19. Marx, *1844 Manuscripts*, 297.

20. Marx, *Capital*, 1:283.

21. Ibid., 1:283–84.

22. Ibid., 1:129. Marx adds the qualifier "socially necessary" to account for the fact that if, for example, one worker produces a table in five hours while another produces a table of the same quality in ten hours, the second worker's additional five hours of labor obviously add nothing to the value of his table.

23. Marx, *Grundrisse*, 296.

24. Marx, *Capital*, 1:340–417.

25. Marx, *Grundrisse*, 294.

26. Marx and Engels, *Communist Manifesto*, 497.

27. Marx, *1844 Manuscripts*, 273.

28. Ibid., 313.

29. Marx, preface to *Contribution to the Critique of Political Economy*, 263.

30. Ibid.

31. Marx, quoted in McLellan, *Karl Marx*, 242.

32. Marx and Engels, *Communist Manifesto*, 519. This passage is especially significant in light of the fact that in an earlier draft of the document by Engels, England rather than Germany had been held up as the most likely location for communist revolution—but Marx revised this in the final version. See Carver, *Marx and Engels*, 85–86.

33. See Marx, *Capital*, 1:353–67. This entire section of *Capital* contains graphic depictions of the extremes in labor exploitation encountered in early capitalist factories.

34. A fascinating reinterpretation of Marx's concept of exploitation in terms consistent with rational choice theory is given in Roemer, *Free to Lose*.

35. Marx, *Capital*, 3:958–59.

36. Marx, *Grundrisse*, 172–73.

37. Marx, *Capital*, 1:342.

38. Marx, *Communist Manifesto*, 499.

39. Marx, *Grundrisse*, 708.

40. Ibid., 712.

41. Karl Marx, "Critique of the Gotha Programme," in *MECW*, vol. 24 (1989), p. 85.

42. Ibid., 87.

43. Marx, *1844 Manuscripts*, 297.

44. Marx and Engels, *The German Ideology*, 47.

45. Ibid., 53.

46. Marx and Engels, *Communist Manifesto*, 494.

47. Levine, *The Tragic Deception*.

48. The best study of the Marx-Engels relationship is Carver, *Marx and Engels*. See also Walicki, *Marxism*, 111–207.

49. Friedrich Engels, *Anti-Duhring: Herr Eugen Duhring's Revolution in Science*, in *MECW*, vol. 25 (1987), p. 112.

50. Ibid., 48–49.

51. Ibid., 49.

52. Ibid., 105.

53. Thus, the bureaucratic corruption of socialist ideals in the name of orthodoxy observed during the Brezhnev era can be traced theoretically to a much earlier stage in the development of Marxism. See Ken Jowitt, "Soviet Neotraditionalism," in *New World Disorder*.

54. These articles later became the basis of Bernstein's book *The Premises of Socialism and the Tasks of Social Democracy* (1899), which has been published in English in an abridged form as *Evolutionary Socialism*.

55. It is telling, in this connection, that Bernstein, without knowing much about either philosopher, discarded the Hegelian basis of Marxism in favor of a vague Kantianism. The crucial importance of the Hegelian contribution to Marx's conception of time has already been noted. For a general summary of Bernstein's views, see Kolakowski, *Main Currents of Marxism*, 2:98–114.

56. Bernstein, *Evolutionary Socialism*, xxiv.

57. Ibid., 2.

58. Ibid., 8.

59. Ibid., 219.

60. Ibid., 220.

61. See Karl Marx, "On the Hague Congress," in *MECW*, vol. 23 (1988), p. 255.

62. See Marx, "Critique of the Gotha Programme," 95. The Gotha Critique was written only three years after Marx's speech to the Hague Congress cited above.

63. Bernstein, *Evolutionary Socialism*, xxvi.

64. Ibid., 202.

65. Ibid., xxix.

66. Rosa Luxemburg, *Reform or Revolution?*, reprinted in *Rosa Luxemburg Speaks*, 33–90.

67. Quoted in Kolakowski, *Main Currents of Marxism*, 2:77.

68. Rosa Luxemburg, *The Mass Strike, the Political Party, and the Trade Unions*, in *Rosa Luxemburg Speaks*, 188.

69. Ibid., 189.

70. See Rosa Luxemburg, *The Russian Revolution*, in *Rosa Luxemburg Speaks*, 365–95.

71. On Kautsky, see Steenson, *Karl Kautsky*; Pierson, *Marxist Intellectuals*; and Donald, *Marxism and Revolution*.

72. This point has been persuasively emphasized in Donald, *Marxism and Revolution*.

73. Kautsky, "Die Revision des Programms der Sozialdemokratie in Osterreich," *Die neue Zeit* 20, no. 2 (1902), quoted in Donald, *Marxism and Revolution*, 30.

74. Quoted in Steenson, *"Not One Man!,"* 207.

75. Dieter Groh, quoted and translated by Steenson in ibid., 227–28.

CHAPTER FOUR

1. The continuing adherence of scholars to the idea that Marxist ideology had no real impact on Lenin's actions is rather surprising in view of the convincing refutation of this position by Neil Harding, whose two-volume work *Lenin's Political Thought* marks a momentous advance in Lenin scholarship. As Harding has shown through detailed empirical study of Lenin's life and writings, Lenin clearly identified himself as an orthodox Marxist theorist—and behaved in ways consistent with this identification—from the age of nineteen until his death. See Harding, *Lenin's Political Thought*, vols. 1 and 2.

2. Fischer, *Life of Lenin*, 28.

3. Plamenatz, *German Marxism*, xxi–xxii.

4. Schapiro, *Communist Party*, 219.

5. Volkogonov, *Lenin*, xxxvii.

6. Von Laue, *Why Lenin?*, 66.

7. Greenfeld, *Nationalism*, 270. The one piece of textual evidence from Lenin's writings cited in this work simply states Lenin's long-standing position that love of one's own nation can play a useful role in the socialist movement—but only when it is placed in the service of the "internationalist" struggle against imperialism.

8. Pipes, *Russia under Bolshevik Rule*, 502.

9. Skocpol, *States and Social Revolutions*.

10. Rabinowitch, *Bolsheviks Come to Power*.

11. McDaniel, *Autocracy, Capitalism, and Revolution*.

12. Goldstone, "Ideology, Cultural Frameworks, and the Process of Revolution," 405–53.

13. The need to place *What Is to Be Done?* in its historical context as Lenin's contribution to the Marxist orthodoxy developed by Plekhanov and Kautsky has been persuasively emphasized by Harding in *Lenin's Political Thought*. However, Harding's analysis portrays *What Is to Be Done?* as essentially a restatement of the views of these previous thinkers; he thus misses the fundamental novelty of the work seen as a proposal for charismatic-rational political organization built upon orthodox premises.

14. On Plekhanov, see Baron, *Plekhanov*.

15. V. I. Lenin, *What Is to Be Done?*, in *Selected Works*, 1:116f.

16. Ibid., 1:149.

17. Karl Marx, *The Eighteenth Brumaire of Louis Bonaparte*, in *Surveys from Exile*, 189.

18. Lenin, *What Is to Be Done?*, 1:114.

19. Ibid., 1:107–8.

20. Ibid., 1:193.

21. Ibid., 1:187.

22. Ibid., 1:114.

23. Ibid., 1:109.

24. Ibid., 1:114.

25. Ibid., 1:198.

26. Ibid. The passage Lenin places in single quotation marks is a quotation from Vera Zasulich in *Zarya*, no. 2–3, p. 353.

27. Ibid., 1:188.

28. Ibid., 1:189.

29. Ibid., 1:194–5.

30. Ibid., 1:178.

31. Ibid., 1:201.

32. See, for example, *One Step Forward, Two Steps Back, Two Tactics of Social Democracy in the Democratic Revolution*, and, after 1917, *The Proletarian Revolution and the Renegade Kautsky* and *Left-Wing Communism: An Infantile Disorder*, in Lenin, *Selected Works*.

33. Williams, *The Other Bolsheviks*.

34. Lenin, *Materialism and Empirio-Criticism*, 178.

35. Ibid., 187.

36. Ibid., 192.

37. Ibid., 337.

38. Lenin, "Karl Marx: A Brief Biographical Sketch with an Exposition of Marxism," in *Selected Works*, 1:40; ellipsis Lenin's.

39. Birken, "Lenin's Revolution," 613–24.

40. Lenin, *Imperialism: The Highest Stage of Capitalism*, in *Selected Works*, 1:723.

41. Ibid.

42. Ibid., 1:714.

43. Lenin, *The State and Revolution*, in *Selected Works*, 2:238–327.

44. Polan, *Lenin*, 20–27. The analysis that follows is indebted to Polan's work. It was Polan who first pointed out theoretically the connection between the argument of *The State and Revolution* and Lenin's faith in a revolutionary "end of time." My analysis differs from Polan's, however, in emphasizing the rational, time-bound elements of Lenin's work as well as the charismatic, time-transcendent ones. In other words, Polan correctly interprets Lenin's vision as charismatic but fails to see how it is more specifically charismatic-rational. See Polan, *Lenin*, 57–58; on Lenin's "politics for the end of time," see 205–6.

45. Lenin, *State and Revolution*, 2:281.

46. Interestingly enough, the same section of *The State and Revolution* from which the above passage was taken contains one of the few instances in Lenin's work in which he allows for the possibility of inconsistency in the writings of Marx or Engels: "Addicts to hair-splitting criticism, or bourgeois 'exterminators of Marxism,' will perhaps see a contradiction between [Engels'] *recognition* of the 'abolition of the state' and repudiation of this formula as an anarchist one in . . . *Anti-Duhring*." Lenin dismisses this criticism as self-evidently absurd, since Engels could not possibly have been an anarchist. Yet the tension in both Marx's and Engels' texts between the charismatic abolition of the state and the modern utilization of the state is precisely the one that remains unresolved in Lenin's work as well. See ibid., 2:280.

47. Ibid., 2:249.

48. Ibid., 2:284–85. Lenin calls this statement "the most theoretically important Engels makes" in the passage he is discussing.

49. Ibid., 2:271–72.

50. Polan, *Lenin*, 81.

51. Lenin, *State and Revolution*, 2:273. Lenin continues on to say that a proper economic foundation for the proletarian state would be "to organize the *whole* economy on the lines of the postal service"—a statement that indeed turned out to be prophetic of later developments in the Soviet Union, albeit in a way Lenin did not intend.

52. Ibid., 2:310.

53. Ibid., 2:308.

54. Ibid.

55. Ibid., 2:309–10.

56. Ibid., 2:312.

57. Ibid., "Postscript to the First Edition," 2:327.

58. Precisely how social conditions under tsarism and the Provisional Government helped to condition working-class support for Lenin's program in 1917 is a fascinating topic that has been examined in detail by McDaniel, *Autocracy, Capitalism, and Revolution*, and Rabinowitch, *Bolsheviks Come to Power*, among others. Of particular interest is the way in which negative prerevolutionary experiences with autocratic capitalist time discipline might have led workers to identify with the Bolshevik ideal of charismatic-rational time. For example, note the description of the temporal regime in one tsarist factory provided by the socialist worker Kanatchikov: "Officially, our workday was supposed to be eleven hours: but to our misfortune, the 'time problem' was decided not by us but by our inventor-entrepreneur. . . . Hanging on the wall of the workshop was a dusty, fly-blown clock, which, obedient to [the entrepreneur's] will, accelerated its pace at the beginning of the workday and slowed down at the end, thereby transforming our eleven-hour workday into twelve hours." It is easy to see how workers subjected to conditions such as these might imagine that an alternative socialist organization of time would lead to a far more energetic and enthusiastic labor discipline than was possible under autocratic capitalism.

Note also Kanatchikov's interesting misinterpretation of the nature of capitalist clock discipline here. In fact, it would be in the *workers'* interest, rather than the capitalist's, if the clock on the wall were accelerated at the beginning of the day. Kanatchikov is apparently complaining about the sped-up *pace of events* at the beginning of the day and the slowed-down *system of time measurement* at the day's end; but he mistakenly associates the first of these complaints as well as the second with an acceleration of the clock. His description thus betrays a lingering tendency to see time as defined by the unfolding of events themselves. See S. I. Kanatchikov, "From the Story of My Life," in Bonnell, *The Russian Worker*, 70.

59. Kautsky, *Terrorism and Communism*, 188. Kautsky's criticism of this "leap" differed from the Menshevik position, however, in that he did not focus on the impossibility of socialist revolution in a peasant society; in fact, he had himself argued in favor of this before 1914—and been heartily supported by Lenin.

60. Ibid., 161–62.

61. The evolution of the Bolshevik Party from its rather amorphous state in 1917 to a disciplined and centralized organization by 1923 has been traced in detail by Service, *Bolshevik Party in Revolution*. In my view, Service's extremely careful study still underestimates the influence of Lenin's prerevolutionary doctrines of party organization on the institutionalization of the Leninist party-state after 1917.

62. On Soviet Taylorism, see Bailes, "Alexei Gastev," 373–94; Sochor, "Soviet Taylorism Revisited," 246–64; and Beissinger, *Scientific Management*.

63. Beissinger, *Scientific Management*, 22.

64. Lenin, "The Immediate Tasks of the Soviet Government," in *Selected Works*, 2:592–93.

65. Ibid., 2:602.

66. Ibid., 2:603.

67. On the *subbotnik* movement, see Chase, "Voluntarism, Mobilisation, and Coercion," 111–28.

68. Lenin, "A Great Beginning: Heroism of Workers in the Rear," in *Selected Works*, 3:168.

69. Ibid.

70. Ibid., 3:177.

71. Positive references to *subbotniki* can be found, however, in "From the Destruction of the Old Social System to the Creation of the New," written in April 1920, in *Selected Works*; and in Lenin's speech to the Third All-Russia Congress of the Russian Young Communist League, delivered October 2, 1920, also in *Selected Works*. The latter speech contains the interesting prediction that fifteen-year-olds in 1920 "will be living in a communist society in ten or twenty years' time" if the Young Communist League can "teach all young people to engage in conscious and disciplined labor from an early age" (423).

72. Lenin, "Speech to the Eleventh Congress of the R.C.P.(B.)," in *Selected Works*, 3:625.

73. Lenin, quoted in Lewin, *Political Undercurrents*, 86.

74. Lenin, "Better Fewer but Better," in *Selected Works*, 3:715.

75. Ibid.

76. Ibid., 3:718.

77. Ibid., 3:725.

78. Lenin, "On Cooperation," in *Selected Works*, 3:703–4.

79. Stalin's political role in the twenties is discussed in the next chapter.

80. Deutscher, *The Prophet Armed*, 183.

81. Trotsky, quoted in ibid., 159.

82. Richard B. Day has argued that a "myth of Trotskyism" as based on the doctrine of "permanent revolution" has clouded our understanding of Trotsky's political activities from 1917 through 1930. Day argues that Trotsky's economics were actually quite inconsistent, since he argued for isolationist economic policies after the Bolshevik victory in the Civil War and then shifted to advocating integration with the world economy in the late 1920s. What Day does not emphasize, however, is that Trotsky remained consistent about one thing throughout this period—the need to speed up the "tempo" of building socialism. At first, Trotsky thought this could be done by the militarization of Soviet labor alone; later he felt that it would require the support of more advanced capitalist nations—but in neither instance did he feel the revolution had any time to wait. Thus, the ideal of permanent revolution—or charismatic Marxism-Leninism—can be seen in Trotsky's economic theorizing throughout the NEP period. See Day, *Leon Trotsky*.

83. It is true that Trotsky at first had Lenin's support for these experiments, but this coincidence of views was only temporary. Lenin was at this time, as pointed out above, arguing for experimentation with all types of "socialist discipline," including both "communist subbotniki" and "labor armies." But Lenin, unlike Trotsky, quickly abandoned the notion that revolutionary enthusiasm should play the major role in socialist economic construction when the disastrous state of the Russian economy became clear to him in 1920–21.

84. Trotsky's speech to the trade union leaders, January 12, 1920, quoted in Deutscher, *The Prophet Armed*, 493.

85. Trotsky, quoted in ibid., 495.

86. On Preobrazhensky's views, see Erlich, *Soviet Industrialization Debate*.

87. Deutscher, *The Prophet Unarmed*, 112.

88. Trotsky, quoted in ibid., 197.

89. Trotsky, quoted in ibid., 199.

90. Trotsky, quoted in ibid., 139; Deutscher's emphasis. Deutscher stresses that Trotsky actually refused to pronounce his own views incorrect in this speech, remarking, "I cannot say so, comrades, because I do not think so." But the fact that Trotsky felt compelled to preface his defense of his views with such a paean to the party's necessary rightness vividly illustrates how Trotsky's belief in Leninist party discipline constrained his ability to lead an effective opposition to the party center.

91. Consider, for example, the verdict in Carr, *Socialism in One Country*, 1:156–57: "Zinoviev never succeeded in attaining either depth of conviction or depth of understanding; and this innate superficiality, among men who treated the subtleties of doctrine with passionate earnestness, won him an unenviable reputation for shiftiness and lack of scruple. It frequently appeared that there was no principle which he was not prepared to sacrifice on the spur of the moment to the cause of political expediency or personal advancement. When attacked, he quickly abandoned his positions or defended them without courage or dignity."

92. Zinoviev, *Le Leninisme*.

93. Carr, *Socialism in One Country*, 1:157.

94. Ibid., 1:158.

95. Ibid., 1:155. Zinoviev also coined the word *Trotskyism* as a pejorative label.

96. It is significant, in this respect, that Zinoviev's political base was Leningrad. He fiercely opposed the transfer of the capital to Moscow in 1918, preferring to remain as head of party affairs in the "cradle of the revolution" itself. See Carr, *Socialism in One Country*, 154.

97. For an analysis of the establishment of the Lenin cult in the 1920s, see Tumarkin, *Lenin Lives!*

98. Zinoviev, *Le Leninisme*, 264.

99. Ibid., 265.

100. Ibid., 193.

101. Cohen, *Bukharin*, xix.

102. This in itself is a fascinating pattern that we will find repeated again in the case of Malenkov, by all accounts a perfectly faithful disciple of Stalin before his emergence as a proponent of evolutionary socialism shortly after Stalin's death. The explanation for this pattern, I would argue, is the inherent difficulty in being a consistent Marxist in times of revolutionary retrenchment such as the periods Bernstein, Bukharin, and Malenkov lived through. Having first tried to be true to a pure vision of revolutionary action in historical periods that seemed to allow no real scope for revolutionary success, these figures instinctively switched over to the other side of the Marxian dualism and began to advocate time discipline and historical patience rather than charismatic time transcendence.

103. Bukharin's biographer Cohen mentions, but does not really attempt to refute, later Stalinist accusations that Bukharin was a "Soviet Bernstein," calling the analogy an "interesting" one. This is one of the few instances, if not the only one, where Cohen agrees with any aspect of the later Stalinist condemnation of Bukharin and the right (Cohen, *Bukharin*, 134).

104. Interestingly, Bukharin was the only one of the Bolsheviks to make himself at all familiar with the work of Max Weber. Although this demonstrates Bukharin's characteristic affinity with more ideal-typically modern forms of social science, it cannot be said that his understanding of Weber was especially deep. The very starting point of Weber's analysis of politics, after all, is that all political order is inherently based on domination—a presupposition that Bukharin certainly ignored in his early conceptions of socialism.

105. Bukharin, quoted in Cohen, *Bukharin*, 91.

106. Bukharin, quoted in ibid., 94.

107. Cohen, *Bukharin*, 95.

108. Bukharin explicitly subordinated even communism to the laws of time in *Historical Materialism*: "Under communism, man will remain a portion of nature subject to the general law of cause and effect. . . . In a word, the deterministic theory will remain in full force in communist society also" (42).

109. Bukharin, quoted in Cohen, *Bukharin*, 117.

110. Bukharin, quoted in ibid., 147.

111. Fitzpatrick, *The Russian Revolution*.

112. Informative accounts of the battle between Gastev and Kerzhentsev are provided in Bailes, "Alexei Gastev"; Beissinger, *Scientific Management*; Sochor, "Soviet Taylorism Revisited"; and Stites, *Revolutionary Dreams*, 149–58. The writings and political activity of Strumilin have not yet been studied in depth, though they deserve greater attention.

113. Gastev, quoted in Stites, *Revolutionary Dreams*, 151.

114. Gastev, *Kak Nado Rabotat'*, 29–30.

115. Gastev, "Marks i Ford," in ibid, 311.

116. A collection of Kerzhentsev's essays of the 1920s have been reprinted in Kerzhentsev, *Bor'ba Za Vremia*. Trotsky's enthusiasm for Kerzhentsev's "Time League" is mentioned in Carr, *Socialism in One Country*, 1:383 n. 3.

117. Stites, *Revolutionary Dreams*, 157.

118. Ibid., 158.

119. Beissinger, *Scientific Management*.

120. Strumilin, "Ratsionalizatsiia Truda i Sverkhurochnye Raboty," *Vestnik Truda*, no. 1–2 (1921), reprinted in idem, *Problemy Ekonomiki Truda*, 46.

121. Ibid., 51.

122. Ibid., 49.

123. Ibid., 49–50.

124. Strumilin, "Biudzhet Vremeni Russkogo Rabochego V 1922 Gody," *Voprosy Truda*, no. 3–4 (1923), reprinted in idem, *Problemy Ekonomiki Truda*, 190.

125. Ibid.; see also "Biudzhet Vremeni Russkogo Krest'ianina," "Biudzhet Vremeni Rabochikh 1923/24 g.," and "Biudzhet Vremeni Sluzhashchikh," 167–249.

126. On Strumilin's defense of "teleological" planning against the alternative "genetic" approach, see Carr, *Socialism in One Country*, 1:494–501, and *Foundations of a Planned Economy*, Vol. I(2), 787–801.

127. Strumilin, quoted in Carr, *Foundations of a Planned Economy*, vol. 1(2), 789–90.

128. Strumilin, quoted in Cohen, *Bukharin*, 266.

CHAPTER FIVE

1. Stephen Cohen has argued that the term *Leninism* is a misleading one in this context, since it understates the diversity in the views of the various members of Lenin's revolutionary leadership; he proposes the term *Bolshevism* be used instead. I will use the conventional terminology here, however, since all the major participants in the elite debates within the party in the twenties themselves accepted the designation *Leninist*. See Cohen, "Bolshevism and Stalinism," in *Rethinking the Soviet Experience*.

2. There are far too many works detailing these aspects of Stalinism to provide a complete list here. However, see, for example, Medvedev, *Let History Judge*; Solzhenitsyn, *The Gulag Archipelago*; and Conquest, *Harvest of Sorrow*. The precise number of victims of Stalin's policies during collectivization, the Terror, and World War II is a matter of some historical controversy; but that these years in Soviet history were extremely bloody ones, by any comparative standard, is not in doubt.

3. See, for example, Arendt, *Origins of Totalitarianism*.

4. See, for example, Fainsod, *How Russia Is Ruled*.

5. Cohen, "Bolshevism and Stalinism," in *Rethinking the Soviet Experience*, 40.

6. Cohen, "Sovietology as a Vocation," in ibid.

7. Fainsod's classic *Smolensk under Soviet Rule*, for example, provides as nuanced an analysis of the nature of Soviet power in the rural regions of the USSR as anything written since.

8. See, for example, Fitzpatrick, *Education and Social Mobility*; Lewin, *Making of the Soviet System*; and Getty, *Origins of the Great Purges*.

9. Cohen, "Bolshevism and Stalinism."

10. One of the few comparative analyses of Stalinism that emphasizes the role of ideology is presented in Chirot, *Modern Tyrants*.

11. Fainsod, *How Russia Is Ruled*, 102.

12. Carr, *Socialism in One Country*, 1:185.

13. Cohen, *Bukharin*, 224–25.

14. One notable exception to the consensus that Stalin was a theoretical nonentity—leaving aside Stalinist historians themselves, of course—is Tucker, particularly the two volumes of his biography of Stalin, *Stalin as Revolutionary* and *Stalin in Power*. Tucker's analysis has influenced me greatly; however, I disagree strongly with his argument that Stalin was in fact "a Bolshevik of the radical right, who blended his version of Leninist revolutionism with Great Russian nationalism." Not only does this formulation contradict the Marxist conception of the political "right," it seems of limited utility as an explanation of Stalin's activity in the 1920s. Tucker's evidence that Stalin relied on examples from the Russian past, particularly Peter the Great, in formulating his solution to the problems of "socialist construction" is sketchy, drawn mainly from textbooks on Russian history that Stalin *might* have read. The evidence for Stalin's Russian nationalism after 1934 is far more compelling. See *Stalin in Power*, xv, 44–65.

15. Carr, *Socialism in One Country*, 1:184.

16. Stalin, *Marxism*.

17. V. I. Lenin, "Letter to the Congress," in *Selected Works*, 3:680–81.

18. Stalin, quoted in Tucker, *Stalin as Revolutionary*, 366.

19. Roy Medvedev has argued that *The Foundations of Leninism* should not be considered an original work of Stalin's but rather that it was largely cribbed from the

work of F. A. Ksenofontov, whose treatise *Lenin's Doctrine of Revolution* had been sent to Stalin for review shortly before the appearance of Stalin's own pamphlet. Although it is clear that Stalin relied heavily on, and in one or two cases directly copied, Ksenofontov's previous efforts to systematize Leninism, there are nonetheless significant differences between the two works; as Tucker notes, there are no sections in Ksenofontov's book on "strategy and tactics" or "style in work" as there are in Stalin's—and these are precisely the parts of Stalin's pamphlet that are most crucial according to the argument I present here. In addition, many of the similarities between the two works can be attributed to the fact that both follow the words of Lenin himself rather closely. See Medvedev, *Let History Judge*, 508–10; Tucker, *Stalin as Revolutionary*, 324–29.

20. Deutscher, *Stalin*, 271.

21. J. V. Stalin, *The Foundations of Leninism*, in Stalin, *Selected Works*, 16.

22. Ibid., 95–96.

23. Ibid., 98.

24. Ibid., 99.

25. Ibid., 61.

26. Ibid., 86; my emphasis.

27. Ibid., 96.

28. Ibid., 100. Tucker mentions that Stalin based his formulation on an earlier remark of Bukharin's. Unfortunately, I have not been able to trace this reference. However, Tucker also analyzes this concluding section of *The Foundations of Leninism*, in line with the analysis presented here, as evidence that Stalin was "pursuing a purpose beyond that of interpreting Lenin's views," by "implicitly offering the younger party generation an ideal leader-type to emulate and at the same time casting himself as an example of it." See Tucker, *Stalin as Revolutionary*, 318–19. In any case, by using the formulation "American efficiency plus Russian revolutionary sweep" as the basis of the concluding section of his first major independent theoretical work, Stalin was in effect making the notion of the "Leninist style in work" a key component of his own independent theoretical position.

29. Stalin, *Foundations of Leninism*, 100.

30. Stephen Cohen claims that while Stalin was the first to use the *formal* expression "socialism in one country," the "requisite reasoning" was really Bukharin's. However, the argument is rather weak. As Cohen admits, Bukharin never used the phrase explicitly until after Stalin had formulated it; Cohen sees the idea as implicit in earlier statements by Bukharin that a Soviet Union based on peasant cooperatives could "grow into socialism." But the whole point of Stalin's formulation—and the reason it provoked such bitter controversy—was that it made *explicit* the idea of the self-sufficiency of the Soviet Union for building a socialist economy. And for a political elite who took theorizing extremely seriously, it was explicit argumentation, and not implicit assumptions, that counted. See Cohen, *Bukharin*, 147–48.

31. The original version of *The Foundations of Leninism*, in fact, contained an explicit rejection of the possibility of the final victory of socialism in Russia without the help of the proletariat of the West. In a rare (if not unique) case of theoretical self-criticism, Stalin explicitly mentions this in his 1925 pamphlet "Concerning Questions of Leninism," arguing that his earlier formulation was "of some service" in the battle against Trotskyism but that it had to be admitted that the argument against "socialism in one country" in *The Foundations of Leninism* needed to be revised. Subsequent

editions of the 1924 work omitted the offending passage. Stalin, "Concerning Questions of Leninism," in *Selected Works*, 194.

32. Stalin, quoted in Tucker, *Stalin in Power*, 58. This paragraph is based closely on Tucker's account.

33. Stalin, "Once More on the Social Democratic Deviation in Our Party," in *Selected Works*, 228–29.

34. On the horrors of Soviet collectivization, see Conquest, *Harvest of Sorrow*.

35. Stalin, "The Right Danger in the C.P.S.U.(B.)," in *Selected Works*, 295, emphasis Stalin's.

36. Ibid., 297.

37. Ibid., 298–99.

38. Nove, *Economic History*, 147.

39. For an argument along similar lines, see Rutland, *Myth of the Plan*.

40. This was the essential meaning behind the seemingly absurd slogan "Five in four/Five in four/Five in four/*And not in five!*" chanted in Soviet kindergartens during this period. See Tucker, *Stalin in Power*, 96.

41. Stalin, "The Tasks of Business Executives," in *Selected Writings*, 199–200.

42. The treatment of the labor force in the early years of Stalinism has recently become the subject of a rich and detailed scholarly literature. See, for example, Andrle, *Workers in Stalin's Russia*; Filtzer, *Soviet Workers and Stalinist Industrialization*; and Kuromiya, *Stalin's Industrial Revolution*; Siegelbaum, *Stakhanovism*, as well as earlier works such as Conquest, *Industrial Workers*.

43. See Conquest, *Industrial Workers*, 47–73; Andrle, *Workers in Stalin's Russia*, 146–155.

44. Significantly, pure time wages were seldom used in Soviet enterprises; it is hard to reward a worker who is paid a certain number of rubles per hour for compressing activity into a smaller amount of time, since this would actually lower such a worker's wages. The rational thing for an unenthusiastic worker on a time-based salary to do, other things being equal, is to relax. Time wages under early British capitalism did not produce this effect because slacking on the job could lead to unemployment and perhaps starvation. This was not a constraint on the Soviet worker in the Stalinist period, when Soviet industry was faced with a constant massive labor shortage.

45. On "socialist emulation," see Conquest, *Industrial Workers*, 73–77; Siegelbaum, *Stakhanovism*, 40–53; and Kuromiya, *Stalin's Industrial Revolution*, 115–28.

46. On Soviet managers, see the classic work by Berliner, *Factory and Manager*.

47. On the Shakhty affair, see Fitzpatrick, *The Russian Revolution*; Kuromiya, *Stalin's Industrial Revolution*, 12–17.

48. On the mass promotion of the so-called *vydvizhentsy* during the First Five-Year Plan, see Fitzpatrick, *Education and Social Mobility*.

49. See Berliner, *Factory and Manager*, 26, 257–58. That Stalin's socioeconomic order was designed to collapse time itself is evident also in Stalin's introduction of the so-called *nepreryvka*, or "uninterrupted" five-day workweek during the period of the First Five-Year Plan. The *nepreryvka* was designed not only to keep factories running constantly, including during night shifts, but also to eliminate the traditional sacred days of Judaism and Christianity, Saturday and Sunday. Complaints by workers about the impossibility of conducting normal family life under a system in which only 20 percent of the population had any particular day off, combined with increasing resistance to

continuous production at the factory, led to the replacement of the *nepreryvka* with the six-day week, or *chestidnevki*, in 1931. By 1940, Stalin's desire to reach an accommodation with the Russian Orthodox Church, combined with the inability of the party to enforce the *chestidnevki* in the countryside, convinced him to reintroduce the Western calendar with its seven-day week. See Chase and Siegelbaum, "Worktime and Industrialization," and Zerubavel, *Seven-Day Circle.*

50. See, for example, Scott, *Behind the Urals*, and Smith, *I Was a Soviet Worker.*

51. The nature of the actual empirical response of Soviet workers to the Stalinist ideal type of socioeconomic organization is discussed more fully in the Conclusion.

52. The methodological argument I am making here bears a certain resemblance to that of Dunham in her book *In Stalin's Time*, 24–38.

53. The propagandistic nature of *Time, Forward!* is evidenced by the fact that Stalin's February 1931 speech to the managers is cited twice in the book.

54. Kataev, *Time, Forward!*, 220.

55. Ibid., 231.

56. Ibid., 98.

57. Ibid., 235.

58. Ibid., 236.

59. Ibid., 238.

60. Ibid., 166.

61. Ibid., 302–5.

62. Ibid., 314.

63. Ibid., 73.

64. Nove, *Economic History of the USSR*, 192.

65. Fitzpatrick, *Education and Social Mobility.*

66. See Andrle, *Workers in Stalin's Russia*, 146–50; and Filtzer, *Soviet Workers and Stalinist Industrialization*, 96–102, 179–207.

67. Andrle, *Workers in Stalin's Russia*, 54–57.

68. From this perspective, there is an intimate connection between "neotraditionalism" within the Leninist party organization as described by Jowitt and the kind of factory-based neotraditionalism described by Walder in his study of Communist China. See Walder, *Communist Neo-Traditionalism.*

69. The classic description of the dynamics of this process is that of Fainsod, *Smolensk under Soviet Rule.*

70. Stalin, "Report to the Seventeenth Party Congress," in *Selected Works*, 405.

71. Ibid., 406.

72. Ibid., 418.

73. Ibid., 418.

74. Whether or not Stalin himself planned Kirov's murder is, of course, the subject of a long and controversial historical debate. See, for example, Conquest, *The Great Terror*; Tucker, *Stalin in Power*; and Getty, *Origins of the Great Purges.* Although in my opinion the weight of the evidence points to Stalin's guilt, I will not take a position on this controversy here. What Kirov's murder *symbolized* for Stalin—the inadmissibility of ever relaxing in the battle against time—remains the same regardless of which interpretation is correct, and it was this message that was communicated to party cadres after the assassination.

75. On Stakhanovism, see Siegelbaum, *Stakhanovism.*

76. Getty, *Origins of the Great Purges*, 58–92.

77. Nove, *Economic History of the USSR*, 234–35; Siegelbaum, *Stakhanovism*, 179–209.

78. It is possible, though hard to verify, that Sergei Kirov in fact represented such a centrist response to the cultural dilemmas of socioeconomic Stalinism. This would explain why even though Kirov appeared to agree with almost every aspect of Stalin's "General Line" from 1928 to 1934, he might have appeared to the dictator as a severe threat. Kirov, in other words, might have represented the reemergence of Kautskyism and Zinovievism *within* Stalinist socioeconomic institutions themselves—preceding Leonid Brezhnev by thirty years. For Stalin, with his obsessive need to be faithful to what he considered to be revolutionary Leninism by continually struggling against the constraining power of time—even when this meant the simple persistence through time of a "socialist" status quo—this might have been motive enough for murder. I am indebted to George Breslauer for this point.

79. Stalin, "Defects in Party Work and Measures for Liquidating Trotskyites and Other Double-Dealers: Report and Speech in Reply to Debate of the Plenum of the Central Committee of the CPSU(B), March 3, 1937," in *Selected Works*, 435.

80. Ibid., 424.

81. This interpretation of Stalin's motives for the show trials and the "short course" (Stalin, *History of the Communist Party*) is similar to that of Tucker, *Stalin in Power*.

82. This is pointed out in Beissinger, *Scientific Management*, 131: "Nearly every industrial official who in any way had been involved with the NOT [scientific organization of labor] movement in the 1920s and 1930s was systematically eliminated by Stalin between 1935 and 1938." The most striking exception proves the rule here: Stanislav Strumilin, who continued to publish throughout the Stalinist period and beyond.

83. Robert C. Tucker, "The Dictator and Totalitarianism," in *Soviet Political Mind*.

84. This, in essence, was the "Big Deal" of the post-Stalin era so insightfully described by Dunham in *In Stalin's Time*.

85. As is pointed out by Breslauer, *Khrushchev and Brezhnev*, 41.

86. Ibid., 40.

87. Khrushchev, in *Pravda*, February 3, 1955, quoted in Conquest, *Power and Policy*, 255.

88. Khrushchev, quoted in Breslauer, *Khrushchev and Brezhnev*, 55.

89. Breslauer, *Khrushchev and Brezhnev*, 154.

90. The study of time in economics and sociology in the Brezhnev period, too, continued along the basic lines established by Stalinist theorists of the twenties and thirties. Thus, the form of "time-budget analysis" pioneered by Strumilin in the NEP period became one of the standard methodologies of Soviet sociologists of the sixties and seventies—with the same underlying ideological justification. V. D. Patrushev, for example, continued to see the planning of "rational" labor and leisure alike as necessary for the uncovering of limitless economic "reserves": "The detection and elimination of working time wastes due to defects of various types including the social ones, is highly significant to a society, as well as the detection and use of previously undiscovered working-time reserves. . . . Hence, it is an absolute necessity to obtain a rational use of the working and non-working time from all members of a society." See V. D. Patrushev, "Aggregate Time-Balances and Their Meaning for Socioeconomic Planning," in Szalai, *Use of Time*, 431.

Among Soviet economists as well, study of the "time factor" (*faktor vremeni*) remained a central theme. Primary attention continued to be focused on the need somehow to "accelerate" the pace of economic activity in time, with the ultimate standard being the miraculous attainment of instantaneous production: "Under high tempos of conversion of capital expenditures into basic industrial stock, the scale of uncompleted construction is lessened. In the extreme case, where the speed of the turnover of capital investment approaches infinity, its scale declines to zero, and capital investments are, as it were, instantaneously transformed into industrial stock." Krasovskii, *Faktor Vremeni*.

CHAPTER SIX

1. "Refined Stalinist" is the memorable expression used to describe Gorbachev by Vice President Dan Quayle in the early days of the Bush administration.

2. For an insightful analysis of the assumptions underlying the modernization paradigm within American political science in general and Soviet studies in particular, see Janos, *Politics and Paradigms*.

3. For a pre-Gorbachev analysis of Soviet politics in this vein, see Stephen F. Cohen, "The Friends and Foes of Change: Soviet Reformism and Conservatism," in *Rethinking the Soviet Experience*. The assumption that ideology played only a minor role in post-Stalin party politics was quite widespread in Soviet studies; for an early argument of this sort, see Robert C. Tucker, "The Deradicalization of Marxist Movements," in *Marxian Revolutionary Idea*.

4. Hough, *Russia and the West*.

5. Lewin, *The Gorbachev Phenomenon*.

6. The crucial role of generational turnover in understanding Gorbachev has been stressed by Hough, *Russia and the West*, 17–43. However, Hough and I read the experience of the Gorbachev generation very differently. He argues that Gorbachev and his cohort developed an unambiguously pro-Western orientation even during their early years rising through the educational establishment of high Stalinism. Although a detailed critique of Hough's position is beyond the scope of this essay, it seems to me at least doubtful that such influences as Soviet pro-American propaganda during the period of wartime alliance between the United States and the Soviet Union, the reading of prerevolutionary Russian literature in Stalinist secondary schools, and some exposure to classical philosophy in law school were decisive in shaping Gorbachev's cultural outlook, as Hough implies.

7. In *Perestroika*, Gorbachev recounts his impressions of the period of postwar reconstruction in revealing terms:

> In the West they said at that time that Russia would not be able to rise even in a hundred years, that it was out of international politics for a long time ahead because it would focus on healing its wounds somehow. And today they say, some with admiration and others with open hostility, that we are a superpower! We revived and lifted the country on our own, through our own efforts, putting to use the immense potentialities of the socialist system.
>
> And we cannot but mention one more aspect of the matter which is frequently ignored or hushed up in the West, but without which it is simply impossible to

understand us, the Soviet people; along with the economic and social achievements, there was also a new life, there was the enthusiasm of the builders of a new world, an inspiration from things new and unusual, a keen feeling of pride that we alone, unassisted and not for the first time, were raising the country on our shoulders. (41–42)

Later, Gorbachev explicitly compares the period of postwar reconstruction with the present era: "We are living through no ordinary period. People of the older generation are comparing the present revolutionary atmosphere with that of the first few years after the October Revolution or with the times of the Great Patriotic War. But my generation can draw a parallel with the period of the postwar recovery. We are now far more sober and realistic. So the enthusiasm and revolutionary self-sacrifice that increasingly distinguish the political mood of the Soviet people are all the more valuable and fruitful" (68).

8. That Gorbachev took the cultural ideal of disciplined time transcendence seriously on a personal level is evidenced by his being awarded, at the age of eighteen, the Order of the Red Banner of Labor for his work as an assistant to a combine harvest operator. See Doder and Branson, *Gorbachev*, 6.

9. Here I follow Jowitt's definition of political corruption in "Soviet Neotraditionalism."

10. It is crucial to note that the *corruption* of one's faith and the *rejection* of one's faith are two quite different things. Scholarly analysis of the Brezhnev era has too often equated the massive corruption of the party elite with its loss of faith in Marxist-Leninist values. In fact, a corrupt Marxist-Leninist will behave quite differently than a straightforward political opportunist, for the former and not the latter can still feel *shame.* Despite the stagnation of the movement under Brezhnev, the elder members of the Politburo preserved the capacity to feel *ashamed* at the extent of the corruption of Marxist-Leninist revolutionary ideals in the USSR that had taken place under their watch. The selection of Andropov and Gorbachev as general secretaries of the party by the former high priests of Brezhnevism would be inexplicable without taking this into account.

11. The phrase "Leninist Romantic" is aptly used to describe Gorbachev in Jowitt, "Gorbachev: Bolshevik or Menshevik?," in *New World Disorder.*

12. Hough, "Politics of Successful Economic Reform," 6.

13. Gorbachev, "Political Report to the Twenty-seventh Party Congress."

14. Ibid., 32.

15. Ibid., 1. At the same time, Gorbachev implicitly rejected both left and right responses to the Soviet crisis, arguing instead that charismatic "grandeur" and rational "realism" must be combined in party policy. As he put it, "It is our task to conceptualize broadly, in Lenin's style, the times we are living in, and to work out a realistic, thoroughly weighed program of action that will organically blend the grandeur of our aims with the realism of our capabilities."

16. Gorbachev, "Political Report," 9.

17. Urchukin, "Intensivnye Metody Ispol'zovaniia Rabochego Vremeni," 95–96.

18. Shmelev, "Avansy i Dolgi."

19. Shmelev himself, after his article was severely criticized within the CPSU, later retracted his views on unemployment, saying he "got carried away" on this question. See the interview with Shmelev in *Sovetskaya Kul'tura*, translated in *Foreign Broadcast Information Service Daily Report—Soviet Union*, October 18, 1987.

20. Tat'iana I. Zaslavskaia, "Economic Behavior and Economic Development" (1981), in *Voice for Reform*, 5.

21. Zaslavskaia, "The Social Mechanism of the Economy" (1985), in *Voice for Reform*, 59.

22. Zaslavskaia, "Economic Behavior and Economic Development," 13.

23. Zaslavskaia, "Social Justice and the Human Factor" (1986), in *Voice for Reform*, 86.

24. Zaslavskaia, "Urgent Problems in the Theory of Economic Sociology" (1987), in *Voice for Reform*, 123.

25. Ibid., 129.

26. Zaslavskaia, "Creative Activity of the Masses: Social Reserves of Growth" (1986), in *Voice for Reform*, 74.

27. Ibid., 75.

28. The slogan "human factor" is repeated numerous times in Gorbachev's book *Perestroika*. It should be noted that the slogan *perestroika* itself had been used by Gorbachev to a certain extent from the beginning of his tenure. But until the summer of 1986, much more stress was placed on the term *uskorenie*. During the latter half of 1986, the emphasis placed on *perestroika* and its major components—*glasnost'*, democratization, and "new thinking" in foreign policy—began to grow rapidly. January 1987, in my estimation, marks the point by which Gorbachev finally committed himself to the full-scale form of *perestroika* that ultimately destroyed the old Leninist system.

29. Gorbachev, *Perestroika*, 19.

30. Ibid., 50.

31. Ibid., 131.

32. Ibid., 36.

33. Indeed, Gorbachev in *Perestroika* notes with apparent pride that *perestroika* involves "no ready-made formulas" for a reformed USSR. This ideological inhibition against creating new "formulas" for imposing reform from above also accounts for the oft-noted vagueness and ineffectiveness of the Gorbachev program for Soviet economic revitalization, which merely worked to destroy the old Stalinist planning system without building any concrete alternative. See Goldman, "Gorbachev the Economist," 28–44.

34. As Jerry Hough points out, the formal procedure for the selection of the new Congress of People's Deputies, which guaranteed a substantial percentage of seats to members of so-called public organizations—mainly ones subordinate to the Communist Party itself—was in principle far less democratic than the old direct elections to the Supreme Soviet had been on paper. Had Gorbachev wanted to introduce true parliamentarianism in 1988, he could have simply allowed for multiple-candidate elections to the existing Supreme Soviet. See Hough, *Russia and the West*, 205.

35. Although according to the theoretical framework adopted here, Sakharov's evolutionary model of convergence between capitalism and socialism can be seen as a "right" position, Yeltsin's call for the immediate dismantling of centralism as a "left" position, and Ligachev's neo-orthodoxy as a "center" position, the utility of this terminology by 1989 was beginning to break down. Thus Sakharov and Yeltsin became allies—though still disagreeing on fundamental issues—within the Interregional Group of Deputies in the Supreme Soviet, which began to refer to itself as the "left" and to the Ligachev faction as the "right." Rather than invalidating the analysis given

above, such terminological shifts merely illustrate the process of collapse of Marxist-Leninist ideological categories in the late Gorbachev era. Indeed, it could be argued that the formal alliance of Gorbachev's opponents on both the right and the left as a unified bloc against the old party center marked the end of the distinctive Marxist-Leninist tradition of political discourse extending from Marx to Gorbachev.

CONCLUSION

1. Marx and Engels, *Communist Manifesto*, in *MECW*, 6:519.

2. The term *movement regime* has been persuasively applied to the USSR by Robert Tucker. See his "On Revolutionary Mass-Movement Regimes," reprinted in *The Soviet Political Mind*. I use the term *movement society* here to characterize the ideal type that the leadership of the Soviet "movement regime" was striving to realize in practice. Recently, the same term has been used by M. Steven Fish to describe the sort of society that emerged in Russia in the wake of the collapse of Gorbachev's *perestroika*; Fish's usage seems quite consistent with mine in the sense that the revolutionary social conditions surrounding the destruction of Soviet power once again made charismatic leaders and doctrines plausible to a wide range of social audiences after 1989. See Fish, *Democracy from Scratch*.

3. Although this was certainly a difficult standard to live up to, time-budget analyses of the 1920s showed that something close to this ideal of time use was being met in practice surprisingly often among higher-ranking party members. See Zuzanek, *Work and Leisure*.

4. See Siegelbaum, *Stakhanovism*, 210–46.

5. Sheila Fitzpatrick, "Cultural Revolution as Class War," in idem, *Cultural Revolution in Russia*. The sociological composition of the "shock worker" elite has been persuasively detailed in Kuromiya, *Stalin's Industrial Revolution*.

6. Fitzpatrick, "Stalin and the Making of a New Elite."

7. Thompson, "Time, Work-Discipline, and Industrial Capitalism."

8. On the concept of the "new socialist man," see Bauer, *New Man in Soviet Psychology*.

9. The interpretation of storming in Soviet culture that follows is indebted to Ken Jowitt, "Political Culture," in *New World Disorder*.

10. Storming, it should be pointed out, is a characteristic response of traditional cultures to the new temporal demands of industry in the early history of liberal capitalist regimes as well. As E. P. Thompson has shown, the practice of taking Mondays off, and of increasing work activity throughout the week, thereby approximating a cyclical agricultural rhythm in industrial life, was widespread in England until the beginning of the twentieth century. The point, however, is that while from the perspective of the modern conception of time as institutionalized in liberal capitalism such practices were clearly understood as corrupt and eventually rooted out, the temporal norms of Leninism tended instead to endorse storming as a reasonably close approximation to the continuous heroic struggle against time ideally demanded of Soviet workers. The party leadership's stress on "*ritmichnost*'" in economic production did not imply a preference for temporal efficiency over labor heroism. Rather, this meant only that the leadership preferred to see a more constant attainment of the intense activity

characteristically undertaken only at the end of the planning period. See Thompson, "Time, Work-Discipline, and Industrial Capitalism."

11. Gregory Grossman, "Scarce Capital and Soviet Doctrine," 334.

12. This was graphically illustrated when Gorbachev, shortly after his rise to power, immediately began promoting "Stakhanovite" methods in the workplace! See Liaporov, *Stakhanovtsy Vos'midesiatyx*.

13. A vivid depiction of life on the post-Stalin Soviet collective farm is given in Yanov, *Drama of the Soviet 1960s*.

14. Lewin, *The Gorbachev Phenomenon*, 31.

15. The problem of how to characterize the precise nature of cultural norms of work and leisure time use among urban workers in the late Soviet period is a fascinating and complex issue. See, for example, Moskoff, *Labour and Leisure*; Robinson, Andreyenkov, and Patrushev, *The Rhythm of Everyday Life*; Bushnell, "Urban Leisure Culture"; and Gregory, "Productivity, Slack, and Time Theft." On the particular pressures placed on Soviet women's daily time by traditional gender roles, see Lapidus, *Women in Soviet Society*.

16. Kotkin, *Steeltown, USSR*.

17. A related argument is made in Giddens, *Constitution of Society*, 34–37. However, Giddens fails to distinguish adequately between the ideological roots of formal institutionalizations of social time and the cultural roots of informal daily time use.

18. A concise outline of one possible "evolutionary" approach to comparative politics is given in Chirot, *How Societies Change*.

19. Fukuyama, "The End of History?"

Bibliography

Adam, Barbara. *Time and Social Theory*. Oxford: Polity Press, 1990.

Akhundov, Murad D. *Conceptions of Space and Time: Sources, Evolution, Directions.* Translated by Charles Rougle. Cambridge: MIT Press, 1986.

Al-Azm, Sadik J. *Kant's Theory of Time*. New York: Philosophical Library, 1967.

Andrle, Vladimir. *Workers in Stalin's Russia: Industrialization and Social Change in a Planned Economy*. New York: St. Martin's Press, 1988.

Arendt, Hannah. *Between Past and Future: Eight Exercises in Political Thought*. New York: Viking Press, 1968.

——. *The Human Condition*. Chicago: University of Chicago Press, 1958.

——. *The Origins of Totalitarianism*. New York: Harcourt Brace Jovanovich, 1973.

Ariès, Philippe. *Centuries of Childhood: A Social History of Family Life*. Translated by Robert Baldick. New York: Vintage Books, 1962.

——. *The Hour of Our Death*. Translated by Helen Weaver. New York: Vintage Books, 1981.

Avineri, Shlomo. *Hegel's Theory of the Modern State*. Cambridge: Cambridge University Press, 1988.

——. *The Social and Political Thought of Karl Marx*. Cambridge: Cambridge University Press, 1968.

Bailes, Kendall E. "Alexei Gastev and the Controversy over Soviet Taylorism, 1918–1924." *Soviet Studies* 29, no. 3 (1977): 373–94.

——. *Technology and Society under Lenin and Stalin*. Princeton: Princeton University Press, 1978.

Baron, Samuel H. *Plekhanov: The Father of Russian Marxism*. Stanford: Stanford University Press, 1963.

Bauer, Raymond. *The New Man in Soviet Psychology*. Oxford: Oxford University Press, 1952.

Beissinger, Mark. *Scientific Management, Socialist Discipline, and Soviet Power*. Cambridge: Harvard University Press, 1988.

Bell, Daniel. *The Cultural Contradictions of Capitalism*. New York: Basic Books, 1976.

Berliner, Joseph. *Factory and Manager in the USSR*. Cambridge: Harvard University Press, 1957.

Bernstein, Eduard. *Evolutionary Socialism*. Translated by Edith C. Harvey. New York: Schoken Books, 1970.

Birken, Lawrence. "Lenin's Revolution in Time, Space, and Economics and Its Implications: An Analysis of Imperialism." *History of Political Economy* 28 (Winter 1991): 613–24.

Bonnell, Victoria E., ed. *The Russian Worker: Life and Labor under the Tsarist Regime*. Berkeley and Los Angeles: University of California Press, 1983.

Boorstin, Daniel J. *The Discoverers*. New York: Vintage Books, 1983.

Booth, William James. "Economies of Time: On the Idea of Time in Marx's Political Economy." *Political Theory* 19 (February 1991): 7–23.

Breslauer, George. *Khrushchev and Brezhnev as Leaders: Building Authority in Soviet Politics*. London: George, Allen and Unwin, 1982.

Bukharin, Nikolai. *Historical Materialism: A System of Sociology*. New York: International Publishers, 1925.

Bushnell, John. "Urban Leisure Culture in Post-Stalin Russia: Stability as a Social Problem?" In *Soviet Society and Culture: Essays in Honor of Vera S. Dunham*, ed. Terry L. Thompson and Richard Sheldon. Boulder: Westview Press, 1988.

Campbell, Jeremy. *Winston Churchill's Afternon Nap: A Wide-Awake Inquiry into the Human Nature of Time*. New York: Simon and Schuster, 1986.

Carr, E. H. *The Bolshevik Revolution, 1917–1923*. Vol. 2. London: Macmillian, 1952.

——. *Foundations of a Planned Economy, 1926–1929*. Vol. 2. London: Macmillan, 1971.

——. *The Interregnum, 1923–1924*. London: Macmillian, 1954.

——. *Socialism in One Country, 1924–1926*. Vol. 1. London: Macmillan, 1958.

Carr, E. H., and R. W. Davies. *Foundations of a Planned Economy: 1926–1929*. Vol. 1. London: Macmillan, 1969.

Carver, Terrell. *Marx and Engels: The Intellectual Partnership*. Brighton, Sussex: Wheatsheaf Books, 1983.

Chase, William. "Voluntarism, Mobilisation, and Coercion: *Subbotniki* 1919–1921." *Soviet Studies* 41 (January 1989): 111–28.

——. *Workers, Society, and the Soviet State: Labor and Life in Moscow, 1918–1929*. Urbana: University of Illinois Press, 1987.

Chase, William, and Lewis Siegelbaum. "Worktime and Industrialization in the U.S.S.R." In *Worktime and Industrialization: An International History*, ed. Gary Cross. Philadelphia: Temple University Press, 1988.

Chirot, Daniel. *How Societies Change*. Thousand Oaks, Calif.: Pine Forge Press, 1994.

——. *Modern Tyrants: The Power and Prevalence of Evil in Our Age*. New York: Free Press, 1994.

Cohen, Marshall, Thomas Nagel, and Thomas Scanlon, eds. *Marx, Justice, and History*. Princeton: Princeton University Press, 1980.

Cohen, Stephen. *Bukharin and the Bolshevik Revolution: A Political Biography, 1888–1938*. Oxford: Oxford University Press, 1980.

——. *Rethinking the Soviet Experience: Politics and History since 1917*. Oxford: Oxford University Press, 1985.

Conquest, Robert. *The Great Terror: Stalin's Purge of the Thirties*. New York: Macmillan, 1973.

——. *Harvest of Sorrow: Soviet Collectivization and the Terror Famine*. New York: Oxford University Press, 1986.

——. *Industrial Workers in the USSR*. London: Bodley Head, 1967.

——. *Power and Policy in the USSR*. London: Macmillan, 1961.

Day, Richard B. *Leon Trotsky and the Politics of Economic Isolation*. Cambridge: Cambridge University Press, 1973.

De Grazia, Sebastian. *Of Time, Work, and Leisure*. Garden City, N.Y.: Anchor Books, 1964.

Deutscher, Isaac. *The Prophet Armed*. New York: Vintage Books, 1965.

——. *The Prophet Unarmed*. New York: Vintage Books, 1965.

——. *Stalin: A Political Biography*. New York: Vintage Books, 1960.

Dickey, Laurence. *Hegel: Religion, Economics, and the Politics of Spirit, 1770–1807*. Cambridge: Cambridge University Press, 1987.

Doder, Dusko, and Louise Branson. *Gorbachev: Heretic in the Kremlin*. New York: Viking Penguin, 1990.

Donald, Moira. *Marxism and Revolution: Karl Kautsky and the Russian Marxists, 1900–1924*. New Haven: Yale University Press, 1993.

Dunham, Vera. *In Stalin's Time: Middleclass Values in Soviet Fiction*. Cambridge: Cambridge University Press, 1979.

Eliade, Mircea. *Cosmos and History: The Myth of the Eternal Return*. New York: Harper and Row, 1959.

——. *Sacred and Profane*. San Diego: Harcourt Brace Jovanovich, 1987.

Elias, Norbert. *Time: An Essay*. Oxford: Basil Blackwell, 1992.

Elkind, David. *The Hurried Child: Growing Up Too Fast Too Soon*. Reading, Mass.: Addison-Wesley, 1981.

Erlich, Alexander. *The Soviet Industrialization Debate, 1924–1928*. Cambridge: Harvard University Press, 1960.

Evans, Alfred B. *Soviet Marxism-Leninism: The Decline of an Ideology*. Westport, Conn.: Praeger, 1993.

Evans-Pritchard, E. E. *The Nuer*. New York: Oxford University Press, 1982.

Fainsod, Merle. *How Russia Is Ruled*. Cambridge: Harvard University Press, 1967.

——. *Smolensk under Soviet Rule*. Cambridge: Harvard University Press, 1958.

Filtzer, Donald. *Soviet Workers and De-Stalinization: The Consolidation of the Modern System of Soviet Production Relations, 1953–1964*. Cambridge: Cambridge University Press, 1992.

——. *Soviet Workers and Stalinist Industrialization: The Formation of Modern Soviet Production Relations, 1928–1941*. London: Pluto Press, 1986.

——. *Soviet Workers and the Collapse of Perestroika: The Soviet Labour Process and Gorbachev's Reforms, 1985–1991*. Cambridge: Cambridge University Press, 1994.

Fischer, Louis. *The Life of Lenin*. New York: Harper and Row, 1964.

Fish, M. Steven. *Democracy from Scratch: Opposition and Regime in the New Russian Revolution*. Princeton: Princeton University Press, 1995.

Fitzpatrick, Sheila. *Education and Social Mobility*. Cambridge: Cambridge University Press, 1979.

——. *The Russian Revolution*. Oxford: Oxford University Press, 1982.

——. "Stalin and the Making of a New Elite, 1928–1939." *Slavic Review* 38 (September 1979): 377–402.

——, ed. *Cultural Revolution in Russia, 1928–1931*. Bloomington: Indiana University Press, 1984.

Forman, Frieda Johles, and Caoran Sowton, eds. *Taking Our Time: Feminist Perspectives on Temporality*. Oxford: Pergamon Press, 1989.

Fraser, J. T. *The Voices of Time*. 2d ed. Amherst: University of Massachusetts Press, 1981.

Fromm, Erich. *Marx's Concept of Man*. New York: Frederick Ungar, 1966.

Fukuyama, Frances. "The End of History?" *National Interest* 16 (Summer 1989): 3–18.

Gastev, Aleksei K. *Kak Nado Rabotat': Prakticheskoye Vvedenie v Nauku Organizatsii Truda*. Moscow: Ekonomika, 1966.

Getty, J. Arch. *Origins of the Great Purges: The Soviet Communist Party Reconsidered, 1933–1938*. Cambridge: Cambridge University Press, 1985.

Giddens, Anthony. *The Constitution of Society*. Berkeley and Los Angeles: University of California Press, 1984.

Goldman, Marshall. "Gorbachev the Economist." *Foreign Affairs* 69 (Spring 1990): 28–44.

Goldstone, Jack A. "Ideology, Cultural Frameworks, and the Process of Revolution." *Theory and Society* 20 (August 1991): 405–53.

Gorbachev, Mikhail. *Perestroika: New Thinking for Our Country and the World*. New York: Harper and Row, 1987.

——. "Political Report to the Twenty-seventh Party Congress of the CPSU." Translated in *Foreign Broadcast Information Service Daily Report—Soviet Union* (February 26, 1986).

Gordon, A. A., and Klopov, E. V. *Chelovek Posle Raboty: Sotsial'nye Problemy Byta I Vnerabochego Vremeni*. Moscow: Izdatel'stvo Nauka, 1972.

Gould, Stephen Jay. *Time's Arrow, Time's Cycle: Myth and Metaphor in the Discovery of Geological Time*. Cambridge: Harvard University Press, 1987.

Gouldner, Alvin. *The Two Marxisms: Contradictions and Anomalies in the Development of Theory*. New York: Seabury Press, 1980.

Greenfeld, Liah. *Nationalism: Five Roads to Modernity*. Cambridge: Harvard University Press, 1992.

Gregory, Paul R. "Productivity, Slack, and Time Theft in the Soviet Economy." In *Politics, Work, and Daily Life in the USSR: A Survey of Former Soviet Citizens*, ed. James R. Millar. Cambridge: Cambridge University Press, 1987.

Grossman, Gregory. "The Economics of Virtuous Haste." In *Marxism, Central Planning and the Soviet Economy: Economic Essays in Honor of Alexander Erlich*, ed. Padma Desai. Cambridge: MIT Press, 1983.

——. "Scarce Capital and Soviet Doctrine." *Quarterly Journal of Economics* 67 (August 1953): 311–43.

Gunnell, John G. *Political Philosophy and Time: Plato and the Origins of Political Vision*. Chicago: University of Chicago Press, 1987.

Hall, Edward T. *The Dance of Life: The Other Dimension of Time*. Garden City, N.Y.: Anchor Books, 1984.

——. *The Silent Language*. Garden City, N.Y.: Anchor Books, 1981.

Harding, Neil. *Lenin's Political Thought*. 2 vols. New York: St. Martin's Press, 1977.

Hawking, Stephen. *A Brief History of Time: From the Big Bang to Black Holes*. New York: Bantam Books, 1988.

Hegel, G. W. F. *Phenomenology of Spirit*. Translated by A. V. Miller. Oxford: Oxford University Press, 1977.

——. *Philosophy of Right*. Translated by T. M. Knox. Oxford: Oxford University Press, 1967.

——. *Reason in History: A General Introduction to the Philosophy of History*. Translated by Robert S. Hartman. Indianapolis: Bobbs-Merrill Educational Publishing Co., 1953.

Hough, Jerry. "The Politics of Successful Economic Reform." *Soviet Economy* 5 (January–March 1989): 3–46.

——. *Russia and the West: Gorbachev and the Politics of Reform*. New York: Simon and Schuster, 1990.

Inkeles, Alex, and David H. Smith. *Becoming Modern*. Cambridge: Harvard University Press, 1974.

Janos, Andrew. *Politics and Paradigms: Changing Theories of Change in Social Science*. Stanford: Stanford University Press, 1986.

Jowitt, Ken. *New World Disorder: The Leninist Extinction*. Berkeley and Los Angeles: University of California Press, 1992.

Kant, Immanuel. *Critique of Pure Reason*. Translated by Norman Kemp Smith. New York: St. Martin's Press, 1965.

———. *Grounding for the Metaphysics of Morals*. Translated by James W. Ellington. Indianapolis: Hackett, 1981.

———. *Kant's Political Writings*. Edited by Hans Reiss. Translated by H. B. Nisbet. Cambridge: Cambridge University Press, 1970.

Kaplan, Frederick I. *Bolshevik Ideology and the Ethics of Soviet Labor, 1917–1920*. New York: Philosophical Library, 1968.

Kataev, Valentin. *Time, Forward!* Translated by Charles Malamuth. New York: Farrar & Rinehart, 1933.

Kautsky, Karl. *Terrorism and Communism: A Contribution to the Natural History of Revolution*. Translated by W. H. Kerridge. Westport, Conn.: Hyperion Press, 1973.

Kern, Stephen. *The Culture of Time and Space: 1880–1918*. Cambridge: Harvard University Press, 1983.

Kerzhentsev, Platon M. *Bor'ba Za Vremia*. Moscow: Ekonomika, 1965.

Kolakowski, Leszek. *Main Currents of Marxism*. 3 vols. Translated by P. S. Falla. Oxford: Oxford University Press, 1981.

Kornai, Janos. *The Socialist System: The Political Economy of Communism*. Princeton: Princeton University Press, 1992.

Kotkin, Stephen. *Magnetic Mountain: Stalinism as a Civilization*. Berkeley and Los Angeles: University of California Press, 1995.

———. *Steeltown, USSR: Soviet Society in the Gorbachev Era*. Berkeley and Los Angeles: University of California Press, 1991.

Krasovskii, V. P., ed. *Faktor Vremeni v Planovoi Ekonomike*. Moscow: Ekonomika, 1978.

Kuromiya, Hiroaki. *Stalin's Industrial Revolution: Politics and Workers, 1928–1932*. Cambridge: Cambridge University Press, 1988.

Landes, David. *Revolution in Time: Clocks and the Making of the Modern World*. Cambridge: Harvard University Press, 1983.

Lapidus, Gail Warshovsky. *Women in Soviet Society: Equality, Development, and Social Change*. Berkeley and Los Angeles: University of California Press, 1978.

Le Goff, Jacques. *Time, Work, and Culture in the Middle Ages*. Translated by Arthur Goldhammer. Chicago: University of Chicago Press, 1980.

Leibovich, Anna Feldman. *The Russian Concept of Work: Suffering, Drama, and Tradition in Pre- and Post-Revolutionary Russia*. Westport, Conn.: Praeger, 1995.

Lenin, Vladimir I. *Materialism and Empirio-Criticism: Critical Comments on a Reactionary Philosophy*. New York: International Publishers, 1927.

———. *Selected Works*. 3 vols. Moscow: Progress Publishers, 1977.

Levine, Norman. *The Tragic Deception: Marx contra Engels*. Oxford: Clio Books, 1975.

Lewin, Moshe. *The Gorbachev Phenomenon.* Berkeley and Los Angeles: University of California Press, 1988.

——. *The Making of the Soviet System: Essays in the Social History of Interwar Russia.* New York: Pantheon, 1985.

——. *Political Undercurrents in Soviet Economic Debates: From Bukharin to the Modern Reformers.* Princeton: Princeton University Press, 1974.

Liaporov, N. D., ed. *Stakhanovtsy Vos'midesiatyx (o Prodolzhenii i Razvitii Stakhanovtskikh Traditsii).* Moscow: Izdatel'stvo Politicheskoi Literatury, 1987.

Lovell, David W. *From Marx to Lenin: An Evaluation of Marx's Responsibility for Soviet Authoritarianism.* Cambridge: Cambridge University Press, 1984.

Lowenthal, Richard. "Development vs. Utopia in Communist Systems." In *Change in Communist Systems*, ed. Chalmers Johnson. Stanford: Stanford University Press, 1970.

Luke, Timothy W. *Ideology and Soviet Industrialization.* Westport, Conn.: Greenwood Press, 1985.

Lukes, Steven. *Marxism and Morality.* Oxford: Oxford University Press, 1987.

Luxemburg, Rosa. *Rosa Luxemburg Speaks.* Edited by Mary-Alice Waters. New York: Pathfinder Press, 1970.

McDaniel, Tim. *Autocracy, Capitalism, and Revolution in Russia.* Berkeley and Los Angeles: University of California Press, 1984.

MacGregor, David. *The Communist Ideal in Hegel and Marx.* Toronto: University of Toronto Press, 1984.

McLellan, David. *Karl Marx: His Life and Thought.* New York: Harper and Row, 1973.

——. *The Young Hegelians and Karl Marx.* New York: Praeger, 1969.

Maier, Charles S. "The Politics of Time: Changing Paradigms of Collective Time and Private Time in the Modern Era." In *Changing Boundaries of the Political*, ed. Charles S. Maier. Cambridge: Cambridge University Press, 1987.

Malia, Martin. *The Soviet Tragedy: A History of Socialism in Russia, 1917–1991.* New York: Free Press, 1994.

Marx, Karl. *Capital.* Vol. 1. Translated by Ben Fowkes. New York: Vintage Books, 1977.

——. *Capital.* Vol. 3. Translated by David Fernbach. New York: Vintage Books, 1981.

——. *Grundrisse.* Translated by Martin Nicolaus. New York: Vintage Books, 1973.

——. *Surveys from Exile.* Edited by David Fernbach. New York: Vintage Books, 1974.

Marx, Karl, and Frederick Engels. *Karl Marx Frederick Engels Complete Works.* 47 vols. London: Lawrence and Wishart, 1975–93.

Medvedev, Roy A. *Let History Judge: The Origins and Consequences of Stalinism.* Edited by David Joravsky and Georges Haupt. Translated by Colleen Taylor. New York: Alfred A. Knopf, 1972.

Meyer, Alfred G. *Leninism.* New York: Praeger, 1962.

——. *Marxism: The Unity of Theory and Practice.* Cambridge: Harvard University Press, 1970.

Mitford, Jessica. *The American Way of Death.* New York: Simon and Schuster, 1978.

Moskoff, William. *Labour and Leisure in the Soviet Union.* New York: St. Martin's, 1984.

Nisbet, Robert. *Social Change and History: Aspects of the Western Theory of Development*. Oxford: Oxford University Press, 1969.

Nove, Alec. *An Economic History of the USSR*. Harmondsworth, England: Penguin Books, 1984.

O'Malley, Michael. *Keeping Watch: A History of American Time*. New York: Penguin, 1990.

Pierson, Stanley. *Marxist Intellectuals and the Working-Class Mentality in Germany, 1887–1912*. Cambridge: Harvard University Press, 1993.

Pipes, Richard. *Russia under Bolshevik Rule*. New York: Vintage Books, 1991.

Plamenatz, John. *German Marxism and Russian Communism*. New York: Harper and Row, 1965.

Plato. *The Republic*. Translated by Allan Bloom. New York: Basic Books, 1968.

Polan, A. J. *Lenin and the End of Politics*. London: Methuen, 1984.

Popper, Karl. *The Open Society and Its Enemies*. Princeton: Princeton University Press, 1950.

Postone, Moishe. *Time, Labor, and Social Domination: A Reinterpretation of Marx's Critical Theory*. Cambridge: Cambridge University Press, 1993.

Rabinowitch, Alexander. *The Bolsheviks Come to Power: The Revolution of 1917 in Petrograd*. New York: Norton, 1978.

Rawls, John. *A Theory of Justice*. Cambridge: Harvard University Press, 1971.

Rifkin, Jeremy. *Time Wars*. New York: Henry Holt, 1986.

Robinson, John P., Vladimir G. Andreyenkov, and Vasily D. Patrushev. *The Rhythm of Everyday Life: How Soviet and American Citizens Use Time*. Boulder: Westview Press, 1988.

Roemer, John. *Free to Lose: An Introduction to Marxist Economic Philosophy*. Cambridge: Harvard University Press, 1988.

Rutland, Peter. *The Myth of the Plan: Lessons of the Soviet Planning Experience*. London: Hutchinson, 1985.

Scanlan, James P. *Marxism in the USSR: A Critical Survey of Current Soviet Thought*. Ithaca, N.Y.: Cornell University Press, 1985.

Schapiro, Leonard. *The Communist Party of the Soviet Union*. New York: Vintage Books, 1971.

Schor, Juliet. *The Overworked American: The Unexpected Decline of Leisure*. New York: Basic Books, 1991.

Scott, John. *Behind the Urals: An American Worker in Russia's City of Steel*. Edited by Stephen Kotkin. Bloomington: Indiana University Press, 1989.

Service, Robert. *The Bolshevik Party in Revolution: A Study in Organizational Change, 1917–1923*. London: Macmillan, 1979.

Shmelev, Nikolai. "Avansy i Dolgi." *Novy Mir* 6 (June 1987): 142–58.

Siegelbaum, Lewis H. *Soviet State and Society between Revolutions, 1918–1929*. Cambridge: Cambridge University Press, 1992.

———. *Stakhanovism and the Politics of Productivity in the USSR, 1935–1941*. Cambridge: Cambridge University Press, 1988.

Skocpol, Theda. *States and Social Revolutions*. Cambridge: Cambridge University Press, 1979.

Smith, Andrew. *I Was a Soviet Worker*. New York: E. P. Dutton, 1936.

Sochor, Zenovia A. "Soviet Taylorism Revisited." *Soviet Studies* 33, no. 2 (1981): 246–64.

Solzhenitsyn, Aleksandr. *The Gulag Archipelago, 1918–1956: An Experiment in Literary Investigation.* 3 vols. Translated by Thomas P. Whitney. New York: Harper and Row, 1975.

Stalin, Joseph. *History of the Communist Party of the Soviet Union (Bolsheviks): Short Course.* London: Red Star Press, 1973.

———. *Marxism and the National Question.* New York: International Publishers, 1942.

———. *Selected Works.* Tirana, Albania: 8 Nentori, 1979.

———. *Selected Writings.* New York: International Publishers, 1942.

Steenson, Gary P. *Karl Kautsky, 1854–1938: Marxism in the Classical Years.* Pittsburgh: University of Pittsburgh Press, 1978.

———. *Not One Man! Not One Penny!: German Social Democracy, 1863–1914.* Pittsburgh: University of Pittsburgh Press, 1981.

Stites, Richard. *Revolutionary Dreams: Utopian Vision and Experimental Life in the Russian Revolution.* New York: Oxford University Press, 1989.

Strumilin, S. G. *Problemy Ekonomiki Truda.* Moscow: Izdatel'stvo Nauka, 1964.

Szalai, Alexander, ed. *The Use of Time: Daily Activities of Urban and Suburban Populations in Twelve Countries.* The Hague: Mouton, 1972.

Thomas, D. Paul. *Karl Marx and the Anarchists.* London: Routledge and Kegan Paul, 1980.

Thompson, E. P. "Time, Work-Discipline, and Industrial Capitalism." *Past and Present* 38 (1967): 56–97.

Thompson, Terry. *Ideology and Policy: The Political Uses of Doctrine in the Soviet Union.* Boulder: Westview Press, 1989.

Thrift, Nigel. "Owner's Time and Own Time: The Making of a Capitalist Time-Consciousness, 1300–1800." In *Space and Time in Geography: Essays Dedicated to Torsten Hagerstrand,* ed. Allan Pred with Gunnar Tornqvist. Lund, Sweden: CWK Gleerup, 1981.

Toews, John Edward. *Hegelianism: The Path toward Dialectical Humanism, 1805–1841.* Cambridge: Cambridge University Press, 1980.

Trotsky, Leon. *The Revolution Betrayed.* New York: Pathfinder Press, 1972.

Trubnikov, N. N. *Vremia Chelovecheskogo Bytiia.* Moscow: Nauka, 1987.

Tucker, Robert C. *The Marxian Revolutionary Idea.* New York: Norton, 1969.

———. *Philosophy and Myth in Karl Marx.* Cambridge: Cambridge University Press, 1972.

———. *The Soviet Political Mind: Stalinism and Post-Stalin Change.* New York: Norton, 1971.

———. *Stalin as Revolutionary, 1879–1929: A Study in History and Personality.* New York: Norton, 1973.

———. *Stalin in Power: The Revolution from Above, 1928–1941.* New York: Norton 1990.

———, ed. *Stalinism: Essays in Historical Interpretation.* New York: Norton, 1977.

Tumarkin, Nina. *Lenin Lives!: The Lenin Cult in Soviet Russia.* Cambridge: Harvard University Press, 1983.

Urchukin, V. "Intensivnye Metody Ispol'zovaniia Rabochego Vremeni." *Voprosy Ekonomiki* 3 (March 1988): 91–100.

Volkogonov, Dmitri. *Lenin: A New Biography*. Translated by Harold Shukman. New York: Free Press, 1994.

Von Laue, Theodore. *Why Lenin? Why Stalin? Why Gorbachev?: The Rise and Fall of the Soviet System*. New York: Harper Collins, 1993.

Walder, Andrew. *Communist Neo-Traditionalism: Work and Authority in Chinese Industry*. Berkeley and Los Angeles: University of California Press, 1986.

Walicki, Andrzej. *Marxism and the Leap to the Kingdom of Freedom: The Rise and Fall of the Communist Utopia*. Stanford: Stanford University Press, 1995.

Walker, Rachel. "Marxism-Leninism as Discourse: The Politics of the Empty Signifier and the Double-Bind." *British Journal of Political Science* 19 (April 1989): 161–89.

Weber, Max. *Economy and Society*. Edited by Guenther Roth and Claus Wittich. 2 vols. Berkeley and Los Angeles: University of California Press, 1978.

——. *The Protestant Ethic and the Spirit of Capitalism*. Translated by Talcott Parsons. New York: Scribners, 1976.

——. "The Social Psychology of the World Religions." In *From Max Weber: Essays in Sociology*, ed. H. H. Gerth and C. Wright Mills. New York: Oxford University Press, 1946.

Williams, Robert. *The Other Bolsheviks*. Bloomington: Indiana University Press, 1986.

Wolfe, Bertram D. *Three Who Made a Revolution*. New York: Stein and Day, 1984.

Yanov, Alexander. *The Drama of the Soviet 1960s: A Lost Reform*. Translated by Stephen P. Dunn. Berkeley: Institute of International Studies, University of California, Berkeley, 1984.

Zaslavskaia, Tat'iana I. *A Voice for Reform: Essays by Tat'iana I. Zaslavskaia*. Translated and edited by Murray Yanowitch. Armonk, N.Y.: M. E. Sharpe, 1989.

Zerubavel, Eviatar. *Hidden Rhythms: Schedules and Calendars in Social Life*. Berkeley and Los Angeles: University of California Press, 1981.

——. *The Seven-Day Cycle: The History and Meaning of the Week*. New York: Free Press, 1985.

Zinoviev, G. *Le Leninisme: Introduction a l'Etude du Leninisme*. Paris: Bureau d'éditions, de diffusion et de publicité, 1926.

Zuzanek, Jiri. *Work and Leisure in the USSR: A Time-Budget Analysis*. New York: Praeger, 1980.

Index

Napoleon I (emperor of France), 202

Napoleonic Wars, 33

Nazism, 13, 129, 178

Neotraditionalism, 61, 73, 123, 163–64,
178–79, 184, 187–88, 190, 201–2, 211–12,
235 (n. 68)

Nepreryvka, 234–35 (n. 49)

New Economic Policy (NEP), ix, 73, 100–
101, 110–12, 115, 117–18, 122–23, 130–31,
134, 140–41, 144–50, 154, 164, 172, 174,
178

"New thinking," 195–97

Newton, Sir Isaac, 23–24, 28, 57, 59, 62,
83–84

Nicholas II (emperor of Russia), 72

North Korea, xii

Nove, Alec, 152

Novosibirsk, 192

Nuer, 2

October Revolution. *See* Bolshevik
Revolution

Ottoman Empire, 202

Pamyat', 196

Paris Commune, 90, 95

Parsons, Talcott, 5

Paul, Saint, 16

Perestroika, xi, 20, 179–82, 194–200, 210,
212, 239 (nn. 28, 33), 240 (n. 2)

Peter I, ("the Great," tsar of Russia), 232
(n. 14)

Pipes, Richard, 70

Plamenatz, John, 69

Platform of the Forty-six, 111

Plato, 221 (n. 1)

Plekhanov, Georgi, 74–75, 81, 226 (n. 13)

Polan, A. J., 88, 90, 227 (n. 44)

Poland, 197

Prague Spring, 177

Preobrazhensky, Evgenii, 110

Proletkult, 125

Protestantism, 14, 16–18, 23, 31, 33, 37, 60,
207

Provisional Government, 71, 85–86, 94, 228
(n. 58)

Prussia, 31, 33–35, 60

Pugo, Boris, 198–99

Puritanism. *See* Calvinism

Quayle, Dan, 237 (n. 1)

Rabkrin (Workers' and Peasants' Inspection),
100–101

Rational conception of time. *See* Time, modern conception of

Rational-legal domination, 11–12, 73, 118–23,
173, 190, 197, 201–2, 214

Red Army, 99, 108, 130, 198

Revisionism: in Marxist theory. *See* Bernstein, Eduard; Second International

Revisionism: in Soviet historiography,
130–31, 135

Ricardo, David, 45

Roman Catholic Church. *See* Catholicism

Rome, Ancient, 7

Russia, xii, 65, 68, 70–75, 80, 85–87, 94–95,
98, 107–8, 136, 144, 155, 170–71, 188, 194,
198, 200, 237 (n. 7). *See also* Soviet Russia;
Soviet Union

Russian Revolution of 1905, 107

Russian Revolution of 1917, 70–71, 111. *See
also* Bolshevik Revolution

Rykov, Aleksei, 174

Sakharov, Andrei, 197, 239 (n. 35)

Schapiro, Leonard, 69

Scientific organization of labor (NOT),
124–25, 127, 236 (n. 82)

Scotland, 23

Second International, x, 38, 58–59, 61,
64, 66–67, 70, 73–75, 79–80, 85, 87, 94,
104, 106, 113–14, 116, 123, 136, 139, 143,
145, 172, 176, 178, 188–89, 201. *See also*
Marxism

Second World War. *See* World War II

Secret Speech, 175

Shakhty trial, 154

Shmelev, Nikolai, 192, 238 (n. 19)

Shock workers, 154, 163, 166, 205–6, 209,
240 (n. 5)

Smith, Adam, 23, 45